Rare and Interesting Cases in Pulmonary Medicine

Rare and Interesting Cases in Pulmonary Medicine

Ali Ataya, MD
Assistant Clinical Professor of Medicine, Division of Pulmonary
Critical Care, and Sleep Medicine, University of Florida
Gainesville, FL, United States

Eloise Harman, MD
Professor Emeritus of Pulmonary and Critical Care Medicine
University of Florida, College of Medicine, and Staff Physician and
MCU Director, Malcolm Randall VA Medical Center
Gainesville, FL, United States

ACADEMIC PRESS

An imprint of Elsevier
elsevier.com

Academic Press is an imprint of Elsevier
125 London Wall, London EC2Y 5AS, United Kingdom
525 B Street, Suite 1800, San Diego, CA 92101-4495, United States
50 Hampshire Street, 5th Floor, Cambridge, MA 02139, United States
The Boulevard, Langford Lane, Kidlington, Oxford OX5 1GB, United Kingdom

Notices
Knowledge and best practice in this field are constantly changing. As new research and
experience broaden our understanding, changes in research methods, professional practices,
or medical treatment may become necessary.

Practitioners and researchers must always rely on their own experience and knowledge in
evaluating and using any information, methods, compounds, or experiments described herein.
In using such information or methods they should be mindful of their own safety and the safety
of others, including parties for whom they have a professional responsibility.

To the fullest extent of the law, neither the Publisher nor the authors, contributors, or editors,
assume any liability for any injury and/or damage to persons or property as a matter of products
liability, negligence or otherwise, or from any use or operation of any methods, products,
instructions, or ideas contained in the material herein.

Library of Congress Cataloging-in-Publication Data
A catalog record for this book is available from the Library of Congress

British Library Cataloguing-in-Publication Data
A catalogue record for this book is available from the British Library

ISBN: 978-0-12-809590-4

For information on all Academic Press publications visit our website at
https://www.elsevier.com/books-and-journals

  Working together
to grow libraries in
developing countries

www.elsevier.com • www.bookaid.org

Publisher: Mica Haley
Acquisition Editor: Stacy Masucci
Editorial Project Manager: Samuel Young
Production Project Manager: Edward Taylor
Designer: Mark Rogers

Typeset by TNQ Books and Journals

Dedication

To my parents
To my wife
And to all my mentors and patients over the years

– Ali Ataya

To my family
My mentors Dr. Jay Block and Dr. William Bell
And to the patients who have taught me so much

– Eloise Harman

Contents

ORS Contributors   xv
Preface   xvii

Case 1

Pulmonary Amyloidosis   2
Takeaway Points   3
Further Reading   3

Case 2

**Hypocomplementemic Urticarial Vasculitis
    Syndrome**   6
Takeaway Points   6
Further Reading   7

Case 3

Yellow Nail Syndrome   11
Chylothorax   11
Takeaway Points   12
Further Reading   13

Case 4

**Pseudo-Pseudo-Meigs Syndrome
    (Tjalma Syndrome)**   16
Takeaway Points   16
Further Reading   16

Case 5

Thoracic Endometriosis   18
Catamenial Pneumothorax   18
Endometrial Lung Nodules   19
Takeaway Points   19
Further Reading   19

Case 6

Pulmonary Alveolar Proteinosis (PAP)   22
Takeaway Points   23
Further Reading   23

Case 7

Langerhans Cell Histiocytosis   26
Takeaway Points   27
Further Reading   28

Case 8

Swyer-James Syndrome   30
Takeaway Points   30
Further Reading   30

Case 9

Placental Transmogrification of the Lung   33
Takeaway Points   33
Further Reading   33

Case 10

Contarini Syndrome   36
Takeaway Points   36
Further Reading   36

Case 11

Acute Eosinophilic Pneumonia   39
Other Etiologies of Pulmonary Eosinophilia (>10%)   39
Takeaway Points   40
Further Reading   41

Case 12

Benign Metastasizing Leiomyoma   45
Takeaway Points   45
Further Reading   46

Case 13

Birt-Hogg-Dubé Syndrome   48
Takeaway Points   48
Further Reading   49

Case 14

Goodpasture Syndrome   52
Takeaway Points   53
Further Reading   53

Case 15

Complete Tracheal Rings   56
Takeaway Points   56
Further Reading   56

Case 16

Granulomatous-Lymphocytic Interstitial
    Lung Disease   59
Takeaway Points   59
Further Reading   60

Case 17

Antisynthetase Syndrome   62
Takeaway Points   63
Further Reading   63

Case 18

Bilothorax   66
Takeaway Points   66
Further Reading   66

Case 19

Castleman Disease   68
Takeaway Points   69
Further Reading   69

Case 20

Gorham-Stout Syndrome   72
Takeaway Points   72
Further Reading   73

Case 21

Solitary Fibrous Tumor of the Pleura   76
Takeaway Points   77
Further Reading   77

Case 22

Vanishing Lung Syndrome	81
Takeaway Points	81
Further Reading	81

Case 23

Traumatic Pulmonary Pseudocyst	84
Takeaway Points	84
Further Reading	84

Case 24

Acute Fibrinous and Organizing Pneumonia	87
Takeaway Points	87
Further Reading	88

Case 25

Shrinking Lung Syndrome	90
Takeaway Points	91
Further Reading	91

Case 26

Cryoglobulinemia-Associated Diffuse Alveolar Hemorrhage	94
Takeaway Points	94
Further Reading	95

Case 27

Pulmonary Veno-Occlusive Disease	98
Pulmonary Capillary Hemangiomatosis	99
Takeaway Points	100
Further Reading	100

Case 28

Diffuse Idiopathic Pulmonary Neuroendocrine Cell Hyperplasia	102
Takeaway Points	103
Further Reading	103

Case 29

Erdheim-Chester Disease	106
Takeaway Points	107
Further Reading	107

Case 30

Urinothorax   110
Takeaway Points   110
Further Reading   110

Case 31

Hypersensitivity Pneumonitis   112
Takeaway Points   114
Further Reading   114

Case 32

Hermansky-Pudlak Syndrome   116
Takeaway Points   116
Further Reading   117

Case 33

Hughes-Stovin Syndrome   120
Takeaway Points   120
Further Reading   120

Case 34

Bone Cement Implantation Syndrome   122
Takeaway Points   123
Further Reading   123

Case 35

IgG4-Related Systemic Disease   126
Takeaway Points   127
Further Reading   127

Case 36

Idiopathic Pleuroparenchymal Fibroelastosis   131
Takeaway Points   131
Further Reading   132

Case 37

Idiopathic Pulmonary Hemosiderosis   134
Takeaway Points   134
Further Reading   134

Case 38

Kikuchi–Fujimoto Disease	136
Takeaway Points	136
Further Reading	137

Case 39

Lymphangioleiomyomatosis	140
Tuberous Sclerosis Complex	141
Takeaway Points	141
Further Reading	141

Case 40

Pulmonary Light Chain Deposition Disease	144
Takeaway Points	144
Further Reading	145

Case 41

Graft-Versus-Host Disease After Lung Transplantation	148
Takeaway Points	148
Further Reading	148

Case 42

Lipoid Pneumonia	150
Takeaway Points	151
Further Reading	151

Case 43

Pulmonary Lymphomatoid Granulomatosis	155
Takeaway Points	156
Further Reading	156

Case 44

Mounier-Kuhn Syndrome	158
Williams-Campbell Syndrome	159
Takeaway Points	159
Further Reading	159

Case 45

Pulmonary Alveolar Microlithiasis	162
Takeaway Points	163
Further Reading	163

Case 46

Pulmonary Arteriovenous Malformations   166
Hereditary Hemorrhagic Telangiectasia   167
Takeaway Points   167
Further Reading   167

Case 47

Pulmonary Tumor Thrombotic Microangiopathy   170
Takeaway Points   170
Further Reading   171

Case 48

Recurrent Respiratory Papillomatosis   175
Takeaway Points   175
Further Reading   176

Case 49

Diffuse Pulmonary Ossification   178
Takeaway Points   179
Further Reading   179

Case 50

Primary Ciliary Dyskinesia   183
Takeaway Points   184
Further Reading   184

Case 51

Myelomatous Pleural Effusion   186
Takeaway Points   186
Further Reading   187

Case 52

Relapsing Polychondritis   190
Takeaway Points   191
Further Reading   191

Case 53

Rosai-Dorfman Disease   194
Takeaway Points   195
Further Reading   195

Case 54

Silicone Embolism Syndrome   198
Chronic Silicone Embolism Syndrome   198
Takeaway Points   198
Further Reading   198

Case 55

Good Syndrome   200
Other Paraneoplastic/Autoimmune Conditions
    Associated With Thymomas   200
Takeaway Points   201
Further Reading   201

Case 56

Primary Effusion Lymphoma   204
Takeaway Points   204
Further Reading   205

Case 57

Partial Anomalous Pulmonary Venous Return   208
Scimitar Syndrome   208
Pseudo-Scimitar Syndrome   209
Takeaway Points   209
Further Reading   209

Case 58

Chronic Thromboembolic Pulmonary Hypertension   212
Takeaway Points   212
Further Reading   213

Case 59

Congenital Pulmonary Airway Malformation   216
Takeaway Points   217
Further Reading   217

Case 60

Erasmus Syndrome   220
Takeaway Points   220
Further Reading   220

Index   221

ORS Contributors

Humberto E. Trejo Bittar, MD
Assistant Professor, Department of Pathology, University of Pittsburgh Medical Center, Pittsburgh, Pennsylvania

Mitra Mehrad, MD
Assistant Professor, Department of Pathology, Microbiology and Immunology, Vanderbilt University Medical Center, Nashville, Tennessee

Tan-Lucien Mohammed, MD
Associate Professor, Department of Radiology, University of Florida, Gainesville, Florida

Preface

Patients with rare lung disorders are often misdiagnosed or diagnosed late in the course of their disease. This is frequently due to physicians not considering the diagnosis in the first place. However, interest in rare and orphan lung diseases has been increasing among healthcare providers, pharmaceutical companies, and patients.

The goal of this book is to provide an introduction to various rare lung disorders, with the hope that this may assist in the recognition, diagnosis, and treatment of these diseases as they are encountered in clinical practice. This book may also benefit physicians studying for their pulmonary board exams. Each case begins with a case study of a patient with one of these rare diseases followed by a rapid review of the disease or syndrome. This case-centered approach is expected to help the reader to recall the information if they encounter patients with one of these rare diseases.

Ultimately, we hope this book will contribute to improving diagnosis and care of patients with rare lung disorders.

Ali Ataya, MD
Eloise Harman, MD

Case 1

A 60-year-old Caucasian female presents with progressive shortness of breath with exertion and a nonproductive cough for the last year and a half. She is a lifelong nonsmoker, has no significant past medical problems, and is not on any medications.

Examination of the heart and lungs is normal and there is no digital clubbing. A chest computed tomography scan revealed multiple small peripheral nodular opacities in the right upper and lower lobes as well as hilar and mediastinal adenopathy (Fig. 1.1). Endobronchial ultrasound with transbronchial needle aspiration of the mediastinal lymph nodes was performed. Histology is shown in Fig. 1.2. Further workup showed no other organ involvement of the disease.

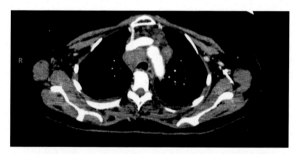

FIGURE 1.1   Chest computed tomography scan with contrast showing enlarged mediastinal 4R node.

FIGURE 1.2   Histology showing clumps of amorphous material with Congo red stain under polarized light.

What is the diagnosis?

Rare and Interesting Cases in Pulmonary Medicine.

PULMONARY AMYLOIDOSIS

Amyloidosis is a systemic disease characterized by extracellular deposition of amyloid, which constitute insoluble β-pleated protein sheets, in different organs. Amyloidosis can be primary/idiopathic (AL type), or secondary/reactive (AA type). The secondary form may occur in the setting of an underlying malignancy, chronic inflammatory, or infectious disease, appear in the setting of chronic renal disease, or be heritable. Isolated pulmonary amyloidosis usually occurs in the setting of the idiopathic form of the disease. Isolated pulmonary amyloidosis is characterized by the occurrence of amyloidosis in the lungs without any systemic involvement.

Patients have nonspecific symptoms due to the diversity of its pulmonary manifestations and tissue biopsy is necessary to make the diagnosis. Isolated pulmonary amyloidosis comes in multiple forms:

1. Tracheobronchial amyloidosis: Most common form. Patients may present with cough, dyspnea, wheezing, or hemoptysis. Patients may have thickened trachea with stenosis. If proximal lesions are present, they may result in fixed upper airway obstruction.
2. Nodular form: Patients may be asymptomatic or present with a cough. A single nodule or multiple small nodular lesions may appear peripherally in the lower lobes. Amyloid nodules may be calcified and cavitate in 10% of cases.
3. Amyloid adenopathy: Amyloid is deposited in the hilar and mediastinal lymph nodes, usually bilaterally. This form of the disease rarely occurs alone or without systemic involvement.
4. Diffuse interstitial form: This is the rarest form of the disease. Amyloid gets deposited in the pulmonary interstitium between the alveoli and blood vessels, impairing gas transfer. Imaging will show a reticular or reticulonodular pattern that may present asymmetrically. Patients succumb to respiratory failure.

Tissue biopsy is the gold standard for diagnosis. Histology will show pink amorphous material that under polarized light will stain apple-green birefringence with Congo red stain.

There is no effective treatment for the disease. Patients with tracheobronchial involvement may undergo bronchoscopic treatment with Nd–YAG laser or clipping for obstructing lesions. For other forms external beam radiation and systemic immunosuppression have been used to halt progression.

This patient underwent further workup that showed no systemic involvement, including a bone marrow biopsy. She was diagnosed with nodular amyloid with hilar and mediastinal lymph node involvement and referred for systemic chemotherapy treatment.

TAKEAWAY POINTS

- Tissue Congo red staining demonstrating apple-green birefringence is pathognomonic for amyloidosis.
- Pulmonary amyloidosis may present as tracheobronchial involvement, nodular disease, thoracic adenopathy, and/or diffuse parenchymal involvement.

FURTHER READING

Thompson, P.J., Citron, K.M., 1983. Amyloid and the lower respiratory tract. Thorax 38, 84–87.
Utz, J.P., Swensen, S.J., Gertz, M.A., 1996. Pulmonary amyloidosis: the Mayo Clinic experience from 1980 to 1993. Ann. Intern. Med. 124, 407–413.

Case 2

A 40-year-old female ex-smoker, with less than a 10 pack-year smoking history, is seen for a 4-year history of exertional dyspnea, significantly worse over the last few months. On system review, she reports a long history of a persistent urticarial skin rash, arthralgias, and recurrent abdominal pain. She also has experienced multiple episodes of angioedema of unknown etiology, for which she required epinephrine and corticosteroids but never endotracheal intubation.

Examination reveals a female with a normal body mass index sitting in the tripod position, saturating 90% on a 3-L nasal cannula oxygen. She has decreased air entry and expiratory wheezes best heard in the lung bases. There are no active skin lesions and no other clinical findings.

Rheumatologic workup shows normal antinuclear antibodies and other autoimmune antibodies are negative. She did have low C3 (27 mg/dL) and C4 (5 mg/dL) levels. She tested negative for C1 esterase inhibitor. Pulmonary function tests (PFTs) revealed FVC 50%, FEV1 24%, TLC 84%, RV 150%, and DLCO 15%, with no response to albuterol. A chest computed tomography scan is shown in Fig. 2.1. Patient's α1-antitrypsin genotype was normal (MM).

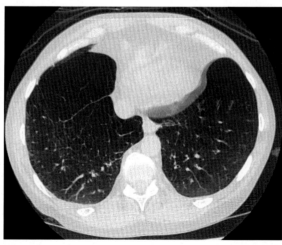

FIGURE 2.1   Chest computed tomography scan of the lung bases showing significant emphysema.

What is the diagnosis?
What are other possible causes of nonsmoking-related emphysema?

HYPOCOMPLEMENTEMIC URTICARIAL VASCULITIS SYNDROME

The constellation of symptoms of urticaria, recurrent angioedema, low complement levels, and basilar emphysema points toward the clinical diagnosis of hypocomplementemic urticarial vasculitis syndrome (HUVS). HUVS is an immune complex-mediated vasculitic disorder that occurs more commonly in females. It can involve multiple other organs with some patients experiencing peripheral neuropathy, nephropathy, recurrent abdominal pains, uveitis, and arthritis. Pulmonary involvement occurs in almost 50% of cases. Most patients may already be established smokers, but the degree of lung disease is out of proportion to the smoking history with basilar predominant emphysema.

Patients will have very low C3 and C4 complement levels and the C1q precipitin antibody will test positive. If the urticarial rash is biopsied, a leukocytoclastic vasculitis pattern is seen with complement deposition at the dermal–epidermal junction. A variant of this syndrome may occur with normal complements levels, termed normocomplementic urticarial vasculitis syndrome.

Immunosuppressive therapy may be used for the skin and nonpulmonary manifestations, but the lung disease progression is rapid and fatal without lung transplant.

Our patients' PFTs showed a very severe obstructive ventilatory defect. She was started on inhaler therapy and oxygen and referred for evaluation for lung transplantation.

Other conditions to consider when dealing with emphysema and bullae in minimal or never smokers may include:

- α1-Antitrypsin deficiency
- Marfans and Ehlers-Danlos syndrome
- Intravenous drug use (cocaine and heroin result in an apical distribution while methylphenidate and methadone result in a basilar distribution)
- Heavy metal exposure to cadmium or indium
- HIV
- Malnutrition
- Recurrent episodes of diffuse alveolar hemorrhage

TAKEAWAY POINTS

- Inquire about urticarial rash and angioedema history in young patients with no risk factors for emphysema.
- Patients with HUVS may be misdiagnosed as systemic lupus erythematosus (SLE); remember that emphysema is not a feature of SLE while dsDNA is negative in HUVS.

FURTHER READING

Lee, P., Gildea, T., Stoller, J., 2002. Emphysema in non smokers: alpha1-antitrypsin deficiency and other causes. Cleve Clin. J. Med. 69, 928–929, 933, 936.

Schwartz, H.R., McDuffie, F.C., Black, L.F., Schroeter, A.L., Conn, D.L., 1982. Hypocomplementemic urticarial vasculitis: association with chronic obstructive pulmonary disease. Mayo Clin. Proc. 57 (4), 231–238.

Case 3

A 42-year-old female is seen for right-sided chest discomfort, worse with inspiration, and mild shortness of breath on exertion for the last few weeks. On further questioning, she states that she has been experiencing sinus drainage for as long as she can remember, as well as a daily cough productive of yellow sputum for many years. She denies any fevers, chills, weight loss, or hemoptysis.

Examination reveals dullness to percussion, decreased tactile fremitus, and decreased breath sounds over the right lower lung base. Extremities show thickened yellow nails (Figs. 3.1 and 3.2) and nonpitting edema of the ankles. A chest X-ray confirms a moderate right pleural effusion and a smaller left-sided effusion.

A thoracentesis is performed revealing a milky appearing, lymphocytic predominant pleural effusion (Fig. 3.3). Pleural fluid triglyceride level is 243 mg/dL, and chylomicrons are present.

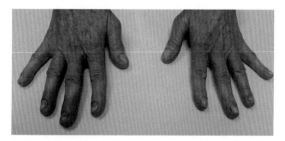

FIGURE 3.1   Thick yellow fingernails.

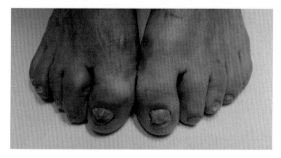

FIGURE 3.2   Thick yellow toenails.

FIGURE 3.3   Milky-white pleural fluid.

What is the diagnosis?
What are the different etiologies of chylothorax and pseudochylothorax?

YELLOW NAIL SYNDROME

Yellow nail syndrome, first described by Samman and White in 1964, is an acquired systemic disease characterized by the development of thick yellow nails, pulmonary manifestations, and lymphedema. Diagnosis requires the presence of two manifestations of the syndrome. The pathogenesis of the disease remains unclear but association with underlying lymphatic flow disruption has been suggested based on abnormal lymphoscintigraphy findings in some patients with the disease. There have also been a few reports of elevated titanium levels in the nails of these patients, but this is of unclear significance.

Pulmonary manifestations include chronic sinusitis, chronic cough, pleural effusions, and bronchiectasis. Bilateral lower airway bronchiectasis is the most common respiratory manifestation of the disease, presenting in almost 46% of cases. Pleural effusions are usually bilateral and exudative by Light's criteria with a predominance of lymphocytes. Chylous effusions appear in 30% of cases. The lymphedema is nonpitting and symmetrical in nature and can involve the upper and low extremities as well as the abdomen. Pericardial effusions have also been reported to occur.

Yellow nail syndrome has also been reported in association with connective tissue diseases, malignancy, immunodeficiency conditions, protein-losing enteropathy, and with use of various drugs.

Management is nonspecific and is directed at the various symptoms of the disease. Sinus symptoms are managed in a standard fashion with decongestants and sinus rinses, and bronchiectasis is managed with airway clearance techniques and antibiotics as needed for infection. Symptomatic pleural effusions should be drained, but often recur. Pleurodesis may be considered in refractory cases. For the treatment of the yellow nails, there have been few reports of successful resolution with the use of vitamin E, zinc, vitamin D3, topical corticosteroids, and clarithromycin therapy.

CHYLOTHORAX

Chylous pleural effusions (chylothorax) appear milky white and opalescent on aspiration, and are characterized by an elevated pleural triglyceride level (>110 mg/dL) and/or the presence of chylomicrons. The effusions are commonly exudative and predominantly lymphocytic on analysis (largely T-lymphocytes), but transudative chylous effusions can occur in up to one-third of cases. Be aware that if the patient is fasting at the time the effusion is sampled, the pleural fluid may have a low triglyceride level and may not appear chylous.

Chylothorax occurs as a result of disruption of lymphatic flow along the thoracic duct or any of its tributaries. It has been documented to occur in the following situations:

- Trauma, which accounts for 50% of cases, and results from thoracic duct injury due to surgery, invasive procedures, or mechanical disruption. Due to the anatomic course of the thoracic duct, injuries above T5 result in left-sided chylothorax, while injuries below T4 result in bilateral or right-sided chylothorax.

- Malignancy, chiefly lymphoma, is the most common nontraumatic cause of chylothorax.
- Superior vena cava thrombosis
- Fibrosing mediastinitis
- Lymphangiomatosis, lymphangiectasias, and lymphangioleiomyomatosis (LAM)
- Thoracic irradiation
- Chylous ascites
- Amyloidosis
- Nephrotic syndrome
- Congenital thoracic duct ectasia
- Down, Turner, and Noonan syndrome
- Gorham-Stout disease (aka. Vanishing bone disease)
- Idiopathic

Management of chylothorax should be focused upon treating the underlying cause. Dietary measures such as a low-fat, high-protein diet, with medium-chained triglycerides supplementation, may decrease the production of chylous fluid. Other potential interventions include thoracic duct ligation or treatment with somatostatin or octreotide. In patients with LAM, treatment with sirolimus may lead to reduction or resolution of chylothoraces.

Pseudochylothorax, also known as chyliform or cholesterol effusion, is less commonly encountered. Pseudochylothorax effusions are usually unilateral. On pleural fluid analysis, the fluid may appear milky white and usually is a neutrophil predominant exudate with high pleural fluid cholesterol (>200 mg/dL) and low triglyceride (<110 mg/dL). Chylomicrons are absent but cholesterol crystals may be present. Unless symptomatic or recurrent, these effusions usually do not require specific treatment.

Pseudochylothorax has been associated with:

- Tuberculosis: This is the most common cause (effusion is usually sterile)
- Rheumatoid arthritis or seronegative arthritis
- Hemothorax
- Malignancy
- Paragonimiasis and echinococcosis
- Trauma

TAKEAWAY POINTS

- Diagnosis of yellow nail syndrome requires the presence of any two of the following: yellow nails, respiratory manifestations, or lymphedema.

- Chylothorax occurs in 30% of cases of yellow nail syndrome and is associated with various other conditions. It is characterized by elevated pleural triglycerides and/or the presence of chylomicrons.
- The majority of chylous effusions are traumatic in nature while malignancy, especially lymphoma, accounts for the majority of nontraumatic cases.
- Pseudochylothorax are effusions with high cholesterol that do not meet pleural analysis criteria for chylothorax.

FURTHER READING

Doerr, C.H., Allen, M.S., Nichols 3rd, F.C., Ryu, J.H., 2005. Etiology of chylothorax in 203 patients. Mayo Clin. Proc. 80 (7), 867–870.

Garcia-Zamalloa, A., Ruiz-Irastorza, G., Aguayo, F.J., Gurrutxaga, N., 1999. Pseudochylothorax. Report of 2 cases and review of the literature. Medicine (Baltimore) 78 (3), 200–207.

Maldonado, F., Ryu, J.H., 2009. Yellow nail syndrome. Curr. Opin. Pulm. Med. 15 (4), 371–375.

Samman, P.D., White, W.F., 1964. The "yellow nail" syndrome. Br. J. Dermatol. 76, 153–157.

Case 4

A 38-year-old female presents to clinic with progressive shortness of breath on exertion for the last 6 weeks, with associated left chest discomfort on inspiration and abdominal fullness. She is a never smoker with a medical history relevant for systemic lupus erythematosus (SLE) and lupus nephritis. Her SLE has been under control for the last few years without any flare-ups.

On examination, there is dullness to percussion and decreased breath sounds over the left hemithorax. Abdominal examination is positive only for abdominal distention with flank dullness and fluid shift consistent with ascites.

A chest X-ray shows a large left-sided pleural effusion. Laboratory testing reveals low complement levels of C3 (20 mg/dL) and C4 (12 mg/dL). A thoracentesis is done and 2 L of straw-colored pleural fluid is aspirated with improvement in the patient's shortness of breath. Pleural fluid is a lymphocytic-predominant transudate, with 68% lymphocytes, 26% monocytes, 6% polymorphs, lactic acid dehydrogenase (LDH) 98 U/L, glucose 92 mg/dL, protein 2.9 g/dL, and pH 7.54 (Serum LDH is 340 U/mL and serum protein is 6.4 g/dL.). Pleural cytology and flow cytometry are unremarkable.

An abdominal ultrasound confirms the presence of ascites and shows normal ovaries. Computed tomography of the abdomen shows ascites without any abdominal or ovarian masses. Echocardiography indicates normal cardiac anatomy and function. Liver function test is within normal limits. Considering the different differential diagnosis, a serum CA-125 level is ordered that comes back elevated at 163 U/mL (normal: 0–25 U/mL).

What is the diagnosis?

PSEUDO-PSEUDO-MEIGS SYNDROME (TJALMA SYNDROME)

There are many potential causes for pleural effusions in SLE including lupus pleuritis, infection, medication, heart failure, nephrotic syndrome, or malignancy such as lymphoma. In our case, the patient had a large unilateral transudative pleural effusion of unclear etiology that did not fit with any of these diagnoses.

Pseudo-pseudo-Meigs syndrome is a rare cause of pleural effusion in patients with SLE that was first described by Tjalma. It has also been reported in a patient with mixed connective tissue disease. This syndrome is characterized by the triad of transudative pleural effusion, ascites, and elevated serum CA-125 levels in the absence of any tumors or ovarian pathology. It is believed that the effusions are the result of a strong inflammatory state, with the elevated serum CA-125 levels resulting from the activation of peritoneal mesothelial cells by various cytokines, such as vascular endothelial growth factor and interleukin-1b. The disease's relationship to lupus activity and lupus nephritis remains unclear.

Pleural effusions, ascites, and benign ovarian fibromas characterize Meigs syndrome, while pseudo-Meigs syndrome refers to pleural effusions and ascites in the setting of a malignant ovarian tumor. CA-125 levels may be normal or elevated in either of these conditions.

The patient was started on cyclophosphamide therapy for her lupus with improvement in both her ascites and pleural effusion. Over the span of few months, her CA-125 levels returned to normal range.

TAKEAWAY POINTS

- Pseudo-pseudo-Meigs syndrome is characterized by pleural effusion, ascites, and elevated serum CA-125 levels in patients with SLE without underlying ovarian pathology.
- Treatment of this condition is focused upon treating the underlying disease.

FURTHER READING

Schmitt, R., Weichert, W., Schneider, W., Luft, F.C., Kettritz, R., 2005. Pseudo-pseudo meigs' syndrome. Lancet 366 (9497), 1672.

Tjalma, W.A., 2005. Ascites, pleural effusion, and CA 125 elevation in an SLE patient, either a Tjalma syndrome or, due to the migrated Filshie clips, a pseudo-Meigs syndrome. Gynecol. Oncol. 97 (1), 288–291.

Case 5

A 23-year-old female complains of coughing up two teaspoons of bright red blood since yesterday. She has not experienced any shortness of breath, fevers, chills, chest pain, or weight change. She does not smoke or use recreational drugs, has no travel or significant exposure history. On further questioning she reports that she has had similar episodes almost monthly for the last 6 months, coinciding with her menstrual periods. She is not on any prescription medications including oral contraceptives, and does not take any over the counter drugs. She is admitted for observation.

Her physical examination is unremarkable and no further hemoptysis is noted. Complete blood count and panel and metabolic profile are unremarkable, chest X-ray is normal. A bronchoscopy is performed and evidence of small amounts of old blood is seen in the right main airway but no active bleeding is noted.

What is the clinical diagnosis?
What are the other manifestations of this disease?

THORACIC ENDOMETRIOSIS

The occurrence of periodic monthly hemoptysis within 72 h before or after menses, in the absence of any other identifiable cause, is consistent with catamenial hemoptysis. Catamenial hemoptysis is part of the spectrum of thoracic endometriosis.

Thoracic endometriosis refers to the implantation of endometrial tissue in the lung parenchyma, pleural surface, or diaphragm. The thorax is the most frequent site of extrapelvic endometriosis. Manifestations of thoracic endometriosis include:

- Catamenial hemoptysis
- Catamenial pneumothorax
- Catamenial hemothorax
- Endometrial lung nodules

The most likely mechanism for the development of thoracic endometriosis is migration of intra-abdominal endometrial implants transdiaphragmatically to the pleural surfaces through congenital or acquired diaphragmatic defects, the so-called "retrograde menstruation." This may explain why endometrial implants on the pleura and diaphragm occur more commonly on the right. However, lung parenchymal endometrial implants are usually bilateral, suggesting lymphatic or hematogenous embolization of endometrial tissue. Another theory is that thoracic endometriosis results from local metaplasia of the coelomic epithelium.

Obtaining a chest computed tomography scan at the time of menses may help reveal parenchymal abnormalities related to the endometrial lung implants that may otherwise be absent at other points during the menstrual cycle. In patients presenting with pneumothorax or hemothorax, video-assisted thoracoscopy may confirm endometrial implants lining the pleura or diaphragm, and allow biopsy confirmation.

Medical treatment with gonadotropin-releasing hormone analogs, or hormonal therapy with estrogen or progesterone, has been used to suppress the bleeding. Airway bleeding may be controlled with argon plasma coagulation. Massive hemoptysis rarely occurs. Successful control with surgical resection or bronchial artery embolization has been described.

CATAMENIAL PNEUMOTHORAX

This is the most common manifestation of thoracic endometriosis (almost 80% of cases). The timing of catamenial pneumothoraces is similar to that of hemoptysis (72 h before or after menstruation). These pneumothoraces occur more commonly on the right, consistent with the higher incidence of endometrial pleural implants in the right hemithorax.

ENDOMETRIAL LUNG NODULES

Parenchymal endometrial implants may present as asymptomatic single or multiple lung nodules. These may also take on the appearance of thin-walled cysts. Bullous disease has also been described. These lesions are usually not associated with any catamenial symptoms.

TAKEAWAY POINTS

- Thoracic endometriosis can manifest as catamenial pneumothorax, hemoptysis, hemothorax, or isolated pulmonary nodules.
- Timing in relation to menstrual cycle in women presenting with recurrent hemoptysis or pneumothorax is an important clue to the diagnosis of thoracic endometriosis.

FURTHER READING

Alifano, M., Roth, T., Broet, S.C., Schussler, O., Magdeleinat, P., Regnard, J.F., 2003. Catamenial pneumothorax: a prospective study. Chest 124 (3), 1004–1008.

Joseph, J., Sahn, S.A., 1996. Thoracic endometriosis syndrome: new observations from an analysis of 110 cases. Am. J. Med. 100 (2), 164–170.

Case 6

A 28-year-old white female with no significant past medical history has had a nonproductive cough for the last 6 months as well as shortness of breath with activity. She denies any chest pain, recent fevers or chills, or wheezing. She has no reflux or sinus symptoms. Her cough and breathlessness have gradually worsened over the last few months, so that she has to pace herself when climbing stairs. She never smoked and has no history of atopy or childhood asthma. Her physical examination is unremarkable with normal breath sounds on auscultation.

Pulmonary function testing showed mild intrinsic restrictive ventilator defect. A chest X-ray (CXR) was done demonstrating bilateral basilar opacities. Given these CXR findings, a computed tomography (CT) scan of the chest was performed (Fig. 6.1).

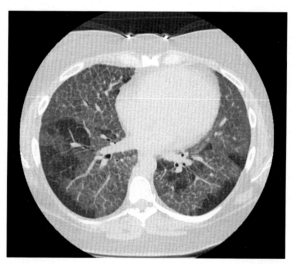

FIGURE 6.1   Chest computed tomography scan showing areas of ground-glass opacity with interstitial thickening.

What is the differential diagnosis given this chest CT pattern?
What are the different treatment options?
What are potential complications associated with this condition?

PULMONARY ALVEOLAR PROTEINOSIS (PAP)

The CT chest findings of patchy ground-glass opacity and superimposed septal thickening are consistent with a pattern called "crazy paving." Although this pattern may be seen in a number of conditions, it is frequently associated with PAP.

Crazy-paving pattern can also be seen in association with:

- Acute respiratory distress syndrome
- Pulmonary edema
- Pneumonia (bacterial, *Pneumocystis jirovecii*)
- Alveolar hemorrhage syndrome
- Adenocarcinoma in situ
- Lymphangitic carcinomatosis
- Nonspecific interstitial pneumonia
- Organizing pneumonia
- Pulmonary veno-occlusive disease
- Lipoid pneumonia
- Alveolar sarcoidosis

This patient underwent bronchoscopy with transbronchial biopsies. Bronchoalveolar lavage (BAL) return was milky in appearance. Transbronchial biopsies showed amorphous eosinophilic material filling the alveoli that was positive on periodic acid Schiff (PAS) staining (Fig. 6.2). These findings confirmed the diagnosis of PAP.

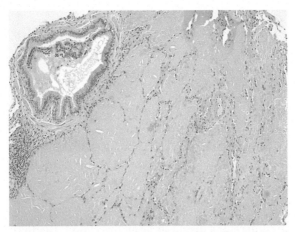

FIGURE 6.2   Lung biopsy showing preserved architecture and alveolar space filling with amorphous eosinophilic granular material. *Courtesy of Humberto Trejo Bittar, MD.*

PAP is a rare lung disease affecting young adults between the ages of 20 and 50 years, and is more common in males. It is characterized by the accumulation of PAS positive lipoproteinaceous material within the alveoli resulting in the ground-glass appearance seen on imaging. Surfactant buildup is believed to be the result of impaired surfactant homeostasis. Dysfunctional macrophage maturation has been associated with impaired surfactant clearance, explaining the association between the development of PAP and underlying hematological malignancies and infections. Most adults with PAP have been found to have autoantibodies directed against granulocyte macrophage colony-stimulating factor (GM-CSF) in both serum and BAL fluid.

The different forms of the disease are:

- Congenital PAP (mutated genes for surfactant proteins B and C)
- Acquired (idiopathic or autoimmune) PAP (85–90% of adult cases)
- Secondary PAP, due to an underlying infection, malignancy (myeloid leukemia), and environmental exposures (silica)

The most common symptoms include progressive shortness of breath and a nonproductive cough, although up to 30% of affected individuals may be asymptomatic. When symptoms of hypoxemia are moderate to severe, whole lung lavage will often provide dramatic symptomatic improvement by removing the lipoproteinaceous material. Whole lung lavage should be performed at a center experienced with the technique. 30 to 40% of patients may require only one lavage. The remaining requires repeated lavages, usually at 6- to 12-month intervals. Some patients with recurrent symptoms may benefit from therapy with GM-CSF. In autoimmune PAP, treatment with GM-CSF therapy (inhaled or systemic) results in symptomatic improvement in half of the treated patients. GM-CSF therapy takes a few months to have its full effect; thus whole lung lavage may still be needed in symptomatic patients. Other potential treatments include lung transplantation, plasmapheresis, and rituximab therapy.

Given the dysfunctional macrophages in PAP, there is increased risk of opportunistic infection, including *P. jirovecii*, mycobacteria, fungi, and *Nocardia* species.

TAKEAWAY POINTS

- Crazy-paving pattern on chest CT imaging can occur in various other conditions besides PAP. BAL and/or transbronchial biopsies are needed to make a diagnosis of PAP.
- Whole lung lavage is the treatment of choice for symptomatic patients; GM-CSF therapy can be used in secondary forms of the disease.
- Patients with PAP are vulnerable to developing opportunistic lung infections with *P. jirovecii*, fungi, mycobacteria, and *Nocardia* species.

FURTHER READING

Seymour, J.F., Presneill, J.J., 2002. Pulmonary alveolar proteinosis. Am. J. Respir. Crit. Care Med. 166, 215–235.
Shah, P.L., Hansell, D., Lawson, P.R., Reid, K.B., Morgan, C., 2000. Pulmonary alveolar proteinosis: clinical aspects and current concepts on pathogenesis. Thorax 55 (1), 67–77.

Case 7

A 32-year-old male presented to the emergency department with acute right-sided chest pain that started 2 h prior to presentation while he was at home watching television. The patient denied any prior history of lung disease or pneumothorax. Past medical history was significant for diabetes insipidus, diagnosed 2 years earlier, for which he was on desmopressin. He was a smoker, but denied use of recreational drugs, occupational or infection exposure or travel history. Family history was unremarkable.

On physical examination, the patient was afebrile and hemodynamically stable. The right chest was hyperresonant to percussion and had diminished breath sounds. Skin examination revealed a papular rash on the arms and legs. A chest X-ray showed a right-sided pneumothorax for which a chest tube was placed. The lung re-expanded, and the chest tube was removed 2 days later. A CT-chest scan of a chest was obtained (Fig. 7.1).

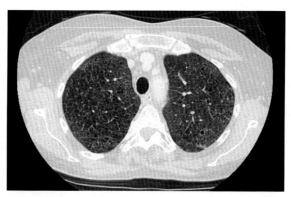

FIGURE 7.1 Computed tomography scan of the chest showing apical irregular cysts bilaterally and scattered centrilobular nodules. These findings were not present in the lung bases.

What is the diagnosis?
What are the systemic manifestations of this disease?

LANGERHANS CELL HISTIOCYTOSIS

The differential diagnosis of cystic lung disease is broad and may include:

- Lymphangioleiomyomatosis
- Langerhans cell histiocytosis (LCH)
- Birt-Hogg-Dubé syndrome
- Lymphocytic interstitial pneumonia
- Amyloidosis
- Light chain deposition disease
- *Pneumocysitis jirovecii* infection
- Cystic metastatic lesions (sarcomas)

The apical distribution of irregular appearing cysts, of different shapes and sizes, and centrilobular nodules in a young male smoker points toward a diagnosis of LCH. The prior history of diabetes insipidus is consistent with this diagnosis. LCH was confirmed by a biopsy of the skin rash, which revealed Langerhans cells with positive staining for CD1a (Fig. 7.2) and S100, confirming the diagnosis of LCH.

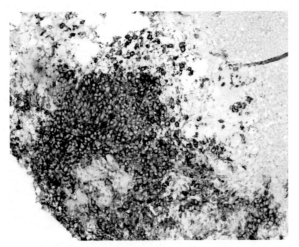

FIGURE 7.2 Skin biopsy staining positive for CD1a cells.

LCH, previously known as histiocytosis X, is a dendritic cell histiocytic disorder characterized by the clonal proliferation of Langerhans cells. LCH usually occurs in both males and females between the ages of 20 and 40, with over 90% of cases occurring among smokers. Pulmonary involvement manifests as multiple centrilobular nodules that transform into cysts of varying sizes and shapes. The cysts appear in the mid to upper lung zones. With progression of the disease, the nodules may completely disappear and cystic involvement may give the lungs an appearance similar to emphysema. Patients are usually asymptomatic, but in advanced cases may complain of shortness of breath and chronic cough. A pneumothorax may be the initial manifestation of the disease in 15% of cases.

Pulmonary function testing may be normal or show a restrictive pattern in early, primarily nodular disease. An obstructive pattern is less common, but may be seen in more advanced cystic disease. Diffusing capacity is frequently reduced out of proportion to other parameters. Pulmonary hypertension has been reported to occur in patients with LCH. Serial echocardiography may be considered to screen for the development of pulmonary hypertension in these patients.

Multiple organs may be involved in LCH, resulting in different clinical manifestations:

- Central nervous system: central diabetes insipidus, headaches, or seizures
- Skin: findings can range from erythematous lesions to ulcerating skin lesions, or a generalized papular rash
- Bone: solitary bone cysts, usually involving the skull, long bones, vertebrae, or ribs
- Lymphatic: lymphadenopathy
- Bone marrow: anemia or pancytopenia

Diagnosis can be established with a biopsy showing Langerhans cells that are positive for S100, CD207, and CD1a. In patients with rash, skin biopsy is recommended as least an invasive way to confirm the diagnosis. Electron microscopy of the LCH may reveal intracytoplasmic inclusions called "Birbeck granules." BRAF-V600E somatic gene mutations have been identified in biopsy samples of almost half of the patients with LCH.

Treatment strategies, according to the Histiocyte Society, depend upon organ system involvement, severity, symptoms, and progression. Smoking cessation is essential. For pulmonary LCH, glucocorticoids have a limited role, primarily in nodular disease. Severe cases may be treated with chemotherapeutic agents, best done via ongoing clinical trials. Lung transplantation is an option in patients with severe pulmonary compromise. The pulmonary hypertension associated with LCH may respond to agents used for pulmonary arterial hypertension such as prostacyclin analogs. Symptomatic bone lesions are managed with radiation therapy and/or surgery. Skin involvement may respond to methotrexate or other agents. Pituitary manifestations such as diabetes insipidus are managed with hormone replacement. Mortality is increased once multi-organ dysfunction takes place, and systemic chemotherapy is often given.

TAKEAWAY POINTS

- Pulmonary LCH is classified as a smoking-associated interstitial lung disease that has a distinct appearance of apical irregular cysts and nodules.

- Biopsy findings are positive for Langerhans cells that stain for CD1a, CD207, and S100. BRAF gene mutations may also be identified in biopsy samples.
- LCH may involve various organs. Multi-organ dysfunction portends a worse prognosis and such patients may require systemic chemotherapy.

FURTHER READING

Gupta, N., Vassallo, R., Wikenheiser-Brokamp, K.A., McCormack, F.X., 2015. Diffuse cystic lung disease. Part I. Am. J. Respir. Crit. Care Med. 191 (12), 1354–1366.

Vassallo, R., Ryu, J.H., Schroeder, D.R., Decker, P.A., Limper, A.H., 2002. Clinical outcomes of pulmonary Langerhans'-cell histiocytosis in adults. N. Engl. J. Med. 346, 484–490.

Case 8

A colleague requests your opinion on the chest computed tomography (CT) findings (Figs. 8.1 and 8.2) of a 52-year-old female. The CT was performed after a chest X-ray (CXR), done for evaluation of a chronic cough, was abnormal. The patient denies any other chronic symptoms, has no medical or surgical history, and is a never smoker.

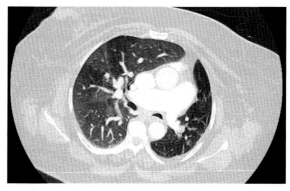

FIGURE 8.1   Computed tomography of the chest showing a small, hyperlucent left lung.

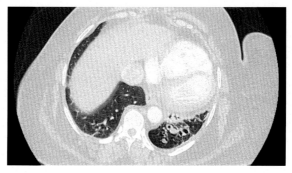

FIGURE 8.2   Computed tomography of the chest findings of bronchiectasis in the left lower lung base.

What is the diagnosis?

Rare and Interesting Cases in Pulmonary Medicine.

SWYER-JAMES SYNDROME

The chest CT findings of a unilateral small, hyperlucent lung with bronchiectasis are typical of Swyer-James Syndrome (SJS), also known as Swyer-James-MacLeod or Brett syndrome.

This acquired syndrome occurs as a result of a postinfectious obliterative bronchiolitis, most commonly associated with adenovirus infection during childhood. Other less commonly associated childhood infections include mycoplasma, influenza, measles, Bordetella, and tuberculosis.

The inflammation and fibrosis resulting from the infection damage the alveolar walls, preventing maturation and growth. Fibrosis and narrowing of the respiratory bronchioles result in air trapping. Overtime, this over distention damages the alveolar capillary beds resulting in reduced pulmonary capillary flow and hypoplasia of the pulmonary arteries. The outcome of these changes is the development of a small, hyperlucent lung. The contralateral lung expands over time to compensate for the smaller affected lung. Occasionally, the affected lung may show changes of bronchiectasis, but this is not necessary to make the diagnosis.

Patients may have a history of a severe or recurrent lung infections during childhood. Most individuals are asymptomatically growing up, but may present in adulthood with recurrent pulmonary infections, such as a productive cough or dyspnea secondary to underlying bronchiectasis. The diagnosis is often made after a CXR is done for an unrelated reason. Pulmonary function tests reveal an obstructive ventilatory defect, while ventilation-perfusion scanning will show a matched ventilation and perfusion defect involving the entire affected lung. Patient management is conservative, aimed at treating and preventing pulmonary infections.

TAKEAWAY POINTS

- SJS manifests as a small, hyperlucent lung that is secondary to childhood postinfectious obliterative bronchiolitis.
- Adenovirus infection is the most common cause of SJS.

FURTHER READING

Lucaya, J., Gartner, S., Garcia-Pena, P., Cobos, N., Roca, I., Linan, S., 1998. Spectrum of manifestations of Swyer-James-MacLeod syndrome. J. Comput. Assist. Tomogr. 22 (4), 592–597.
Swyer, P.R., James, G.C., 1953. A case of unilateral pulmonary emphysema. Thorax 8 (2), 133–136.

Case 9

A 25-year-old female arrived to your clinic complaining of a 1-year history of a nonproductive cough, chest discomfort, and progressive shortness of breath on exertion. She is a never smoker with no significant exposures, no history of drug abuse, and no family history of lung disease. A chest computed tomography scan showed large bullous lesion in the right upper lobe (Fig. 9.1). Routine labs and an alpha-1 antitrypsin level were within normal limits. Given her progressive symptoms, she underwent a right upper lobectomy, which she tolerated well. Her symptoms improved over several months following the surgery.

Microscopic histological analysis of the lung tissue revealed thin-walled bullae containing edematous papillary structures that are composed of proliferative blood and lymphatic vessels, smooth muscle, and adipose tissue, embedded in a villous cellular stroma; a pattern closely resembling placental chorionic villi (Fig. 9.2). These structures were absent in the surrounding normal lung parenchyma.

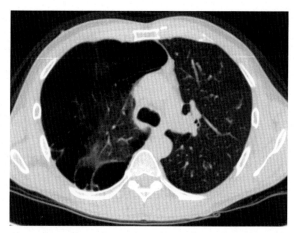

FIGURE 9.1   Chest computed tomography scan showing bullous lesion in the right upper lobe.

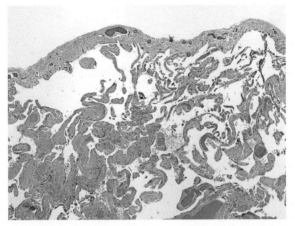

FIGURE 9.2   Lung pathology showing evidence of papillary structures resembling placental villi on a background of emphysematous bullae. *Courtesy of Mitra Mehrad, MD.*

Given the pathology findings, what is the etiology of her bullous lung disease?

PLACENTAL TRANSMOGRIFICATION OF THE LUNG

Placental transmogrification, or placentoid bullous lesion, refers to the histological finding of placental villous structures in the lung parenchyma, most commonly in association with cystic or bullous lesions. It is a rare pathological finding, first described by Chesney in 1979, with less than 30 cases reported in the literature. Placental transmogrification occurs predominantly in males between the ages of 20 and 50 years.

The etiology of this condition, which is found in association with emphysematous and bullous lesions or fibrochondromatous hamartomas, remains unclear. Possible causes include congenital malformation, reaction to underlying emphysema related to lymphatic or vascular abnormalities, or benign proliferation of immature interstitial clear cells with secondary cystic changes.

These lesions manifest, most commonly as unilateral cysts or bullous lung disease, in an otherwise young individual without risk factors for emphysema. There have been reports of it presenting as a pulmonary mass or nodule surrounded by emphysematous changes. Placental transmogrification is almost always benign; however, there is one case report of malignant transformation to a papillary adenocarcinoma.

Affected individuals are usually asymptomatic and may present with incidental findings on imaging. Symptoms associated with this condition include cough, progressive shortness of breath, or in some cases with a pneumothorax as a result of rupture of a bullous lesion.

In all cases, resection of the pathological lung is curative, with good outcomes and no reported recurrence of disease to date.

TAKEAWAY POINTS

- Placental transmogrification of the lung is the histological finding of placental villous structures in the lung.
- It should be considered in the setting of unilateral pulmonary cysts or bullous disease among young adults with no risk factors for emphysema.

FURTHER READING

Cavazza, A., Lantuejoul, S., Sartori, G., Bigiani, N., Maiorana, A., Pasquinelli, G., Rossi, G., 2004. Placental transmogrification of the lung: clinicopathologic, immunohistochemical and molecular study of two cases, with particular emphasis on the interstitial clear cells. Hum. Pathol. 35 (4), 517–521.

McChesney, T., 1979. Placental transmogrification of the lung: a unique case with remarkable histopathologic features. Lab. Invest. 40, 245–246.

Case 10

A 67-year-old patient with a history of heart failure (EF 35%) is seen in clinic for fevers and a productive cough. Chest X-ray shows a right-sided consolidation and bilateral pleural effusions, and a chest computed tomography (CT) scan is performed (Fig. 10.1). Thoracentesis is initially performed on the right and on the left a few days later. The pleural fluid on the right is a serosanguinous exudate, consistent with an uncomplicated parapneumonic effusion. The fluid on the left is straw colored and pleural fluid analysis reveals that it is a transudate. The patient is managed with antibiotics and diuresis, with improvement in symptoms, and resolution of the pleural effusions on follow-up.

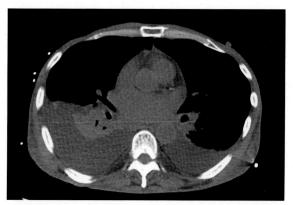

FIGURE 10.1 Bilateral pleural effusions, right larger than left, with associated right lower lobe consolidation.

What is Contarini syndrome?

CONTARINI SYNDROME

Contarini syndrome describes the presence of bilateral pleural effusions that are markedly different in characteristic. Velschius first reported the condition in 1667 when the 95th Dodge of Venice, Francesco Contarini, died with symptoms of foul smelling sputum and orthopnea. On autopsy, he was found to have a large heart, a right-sided effusion; possibly due to heart failure, and pus in the left chest cavity, representing either an empyema or a lung abscess.

Bilateral pleural effusions, when present, are most likely to be similar in characteristic and secondary to the same underlying cause, most commonly associated with heart failure or malignancy. A few reports of bilateral effusions with different characteristics have been described to date, occurring mainly in the setting of heart failure or immunocompromised states. The most common example is a patient with heart failure who develops pneumonia with a parap-neumonic effusion, then decompensated heart failure resulting in the development of a contralateral transudative effusion. In addition, cases of Contarini syndrome have been described with coexisting chylous and transudative effusions, rheumatoid effusions and empyema, and bilateral exudative effusions due to different organisms.

Physicians should be aware of the possibility of this syndrome, specifically, when patients with heart failure also present with fever or a productive cough, when the chest CT scan shows bilateral effusions of different appearance, or when there is only unilateral improvement with treatment. Management is directed toward treating the underlying etiology of each of the effusions.

TAKEAWAY POINTS

- Contarini syndrome is the coexistence of bilateral pleural effusions of different characteristics that are due to different etiologies.
- The most common cause is the precipitation of a heart failure exacerbation in the setting of pneumonia with a parapneumonic effusion.

FURTHER READING

Jarcho, S., 1970. Empyema or hydrothorax in the ninety-five Doge of Venice. Bull. N. Y. Acad. Med. 46, 378–385.

Porcel, J.M., Civit, M.C., Bielsa, S., Light, R.W., 2012. Contarini's syndrome: bilateral pleural effusion, each side from different causes. J. Hosp. Med. 7 (2), 164–165.

Case 11

A 40-year-old male was admitted for progressive shortness of breath, fevers, and a nonproductive cough that started acutely 3 days prior to his presentation. He denies any chest pain, chills, wheezing, or palpitations. He did not have any recent sick contacts but did notice that his symptoms started a day after being exposed to the smoke from burning wood in his backyard. He is a nonsmoker with no significant past medical history, not taking any medications, and no history of recent travel outside the United States.

On admission, he is febrile 38.7°F, heart rate 114 bpm, blood pressure 109/68 mmHg, and saturating 84% on room air. His physical examination was relevant for inspiratory crackles in all lung fields. Patient required intubation and mechanical ventilation for hypoxic respiratory failure and was started with broad-spectrum antibiotics, vancomycin, and cefepime. His chest X-ray (Fig. 11.1) and chest computed tomography (Fig. 11.2) are shown.

Bronchoscopy with bronchoalveolar lavage was performed. The airways were normal on examination; the BAL was cloudy; and cell count showed 46% eosinophils. There was peripheral leukocytosis of 17 k/μL with no peripheral eosinophilia; antinuclear antibody and antineutrophil cytoplasmic antibody (ANCA) testing were negative; and bacterial and fungal cultures and testing for parasites were also negative.

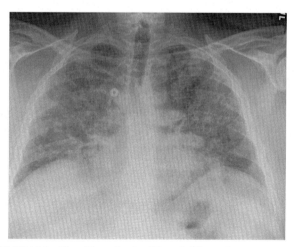

FIGURE 11.1   Chest X-ray showing bilateral diffuse pulmonary opacities.

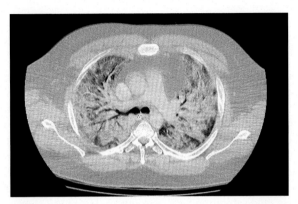

FIGURE 11.2   Diffuse pulmonary parenchymal opacities throughout both lungs on computed tomography chest, most pronounced in the bilateral lung apices.

What is the diagnosis?
What are other causes for pulmonary eosinophilia?

ACUTE EOSINOPHILIC PNEUMONIA

Acute eosinophilic pneumonia (AEP) is a distinct eosinophilic lung disease of unclear etiology that was first described in 1989 in a group of patients presenting with acute hypoxic respiratory failure. It is believed to be as a result of an acute hypersensitivity to an inhaled antigen, most commonly associated with recent onset of cigarette smoking or other smoke inhalation. This hypersensitivity state results in elevated interlukin-5 levels and migration of eosinophils to the lungs producing the clinical manifestations of the disease. Other described manifestations of the disease include exposure to cocaine, tear gas, certain drugs (daptomycin, antidepressants, and progesterone), radiation therapy to the chest, or in conjunction with a connective tissue disease such as rheumatoid arthritis.

Patients often present with a febrile illness, shortness of breath, nonproductive cough, and hypoxic respiratory failure of less than 1-month duration. These symptoms occur within a month (commonly within a week of exposure to the inciting factor). The clinical presentation is initially often mistaken for pneumonia. Imaging of the chest will reveal diffuse bilateral infiltrates and consolidations. BAL will reveal pulmonary eosinophilia (>25%) that cannot be explained by any other disease process. Peripheral eosinophilia is not a feature of this disease. A lung biopsy is not required to make the diagnosis, but when obtained, findings of acute and organizing diffuse alveolar damage are most commonly seen.

Once AEP is identified, treatment involves removal of the inciting factor, supportive care, and corticosteroid therapy. Patients have a good prognosis with a rapid and significant response within a few days of corticosteroid treatment. After resolution of the disease, recurrence rarely occurs.

The recent exposure to the smoke from the burning wood was the inciting factor in this case study. The patient's antibiotics were discontinued once an infectious cause was ruled out; he was treated with corticosteroids with significant clinical improvement. Within 2 days the patient was taken off the ventilator and had resolution of his fevers and leukocytosis. Prednisone was continued for another 2 weeks after the resolution of his symptoms. He did not experience any recurrence.

OTHER ETIOLOGIES OF PULMONARY EOSINOPHILIA (>10%)

- Chronic eosinophilic pneumonia (CEP):
  - Has a less severe presentation than AIP with insidious onset of symptoms of cough and shortness of breath. Patients often report a prior history of asthma. Unlike AIP, peripheral eosinophilia is common. Imaging commonly shows peripheral-based infiltrates that have been described as photographic negative pulmonary edema. There is good response to corticosteroids, but relapses are common.

- Tropical pulmonary eosinophilia:
  - It is characterized by symptoms of wheezing, fever, and eosinophilia (pulmonary and peripheral) in individuals who have recently traveled to, or are organically from, tropical regions. It is associated with the lymphatic filariae *Wuchereria bancrofti* and *Brugia malayi* species. It is the result of an intense immune response to filarial antigens, with high levels of antifilarial antibodies and elevated immunoglobulin E. Patients have good response to the antiparasitic agent diethylcarbamazine, with 20% relapse rates reported.
- Parasitic infections:
  - Löffler syndrome describes pulmonary eosinophilia and migrating pulmonary opacities that occur secondary to helminth larvae traveling through the lungs. It is often secondary to *Ascaris lumbricoides* infection. Other parasitic infections that can present similarly include Paragonimus, Hookworms, Strongyloides, and Toxocara.
- Eosinophilic granulmatosis with polyangiitis:
  - Previously known as Churg Strauss disease. This vasculitis disease manifests with symptoms mimicking asthma and pulmonary eosinophilia. There may be other systemic manifestations of the disease. Measured ANCA is positive in 40–60% of cases.
- Allergic bronchopulmonary mycosis:
  - This hypersensitivity reaction to mycosis species in the lungs was previously described only in the setting of Aspergillus (previously known as ABPA). Diagnosis is based on certain criteria. It occurs in asthmatics or cystic fibrosis patients who present with worsening symptoms.
- Idiopathic hypereosinophilic syndrome (IHES):
  - IHES is a systemic illness of unknown etiology characterized by elevated absolute peripheral eosinophilia on two separate occasions, over a month apart, with biopsy proven eosinophilic organ dysfunction. Up to one-quarter of patients may present with pulmonary symptoms of nonproductive cough or dyspnea. Cardiac involvement carries the highest morbidity and mortality.
- Pulmonary eosinophilia secondary to systemic illnesses:
  - Pulmonary eosinophilia can occur secondary to some bacterial (Mycobacterium) or fungal (Coccidioides) infections, malignancies, acute respiratory distress syndrome, sarcoidosis, and other chronic lung diseases. The degree of pulmonary eosinophilia is usually lower in comparison to the other pulmonary eosinophilic syndromes.

TAKEAWAY POINTS

- AEP is characterized by the acute onset of hypoxic respiratory failure, bilateral pulmonary infiltrates, and pulmonary eosinophilia (>25%), in the absence of peripheral eosinophilia and any other etiology. It is commonly reported to occur after new exposure to smoking. There is good response to corticosteroid, and recurrence is rare.

- CEP is characterized by insidious onset of symptoms, peripheral and pulmonary eosinophilia. Peripheral lung opacities (photographic negative pulmonary edema) are often reported. Patients have good response to corticosteroids, but relapses of the disease are common.

FURTHER READING

Allen, J.N., Pacht, E.R., Gadek, J.E., Davis, W.B., 1989. Acute eosinophilic pneumonia as a reversible cause of noninfectious respiratory failure. N. Engl. J. Med. 321 (9), 569–574.

Cottin, V., Cordier, J.F., 2012. Eosinophilic lung diseases. Immunol. Allergy Clin. North Am. 32 (4), 557–586.

Rhee, C.K., Min, K.H., Yim, N.Y., Lee, J.E., Lee, N.R., Chung, M.P., Jeon, K., 2013. Clinical characteristics and corticosteroid treatment of acute eosinophilic pneumonia. Eur. Respir. J. 41 (2), 402–409.

Case 12

A 44-year-old female is being evaluated for a nonproductive cough and shortness of breath that have been gradually progressing over the last year. Past history is significant for hypertension. She had a hysterectomy when she was in her 30s for a uterine leiomyoma. The patient is a current smoker with a 15 pack-year smoking history. She has no other exposures, and no family history of lung disease. Vital signs and physical examination are normal. A chest X-ray and computed tomography scan of the chest are shown (Figs. 12.1 and 12.2). Pulmonary function tests show an FVC of 94%, FEV1 of 88% of predicted, lung volumes within normal range, and a corrected DLCO of 72%. Video-assisted thoracoscopic surgery (VATS) is done with resection of one of the nodules. Histological evaluation of the lesion is notable for the presence of interlacing bundles of well-defined spindle shaped cells with low mitotic rate (Fig. 12.3). Cells were positive for estrogen and progesterone receptors, and immunohistochemical staining was positive for actin and desmin.

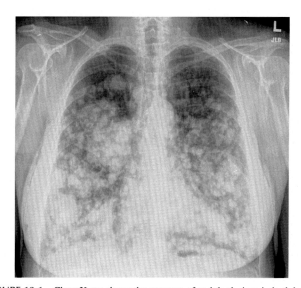

FIGURE 12.1   Chest X-ray shows the presence of nodular lesions in both lungs.

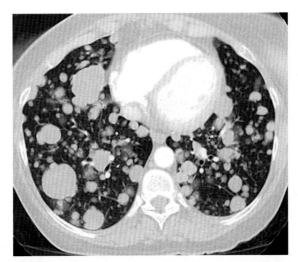

FIGURE 12.2   Computed tomography scan of the chest showing numerous well-circumscribed nodules of various sizes throughout both lungs.

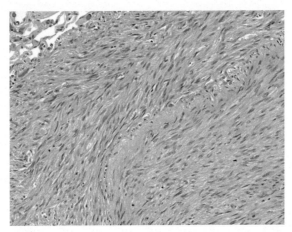

FIGURE 12.3   Fusiform spindle cells with distinctive myogenic features including cigar-shaped nucleus and abundant eosinophilic cytoplasm. No significant atypia or mitotic activity is seen. *Courtesy of Humberto Trejo Bittar, MD.*

What is the diagnosis?

BENIGN METASTASIZING LEIOMYOMA

Uterine leiomyomas are the most common gynecological tumor in premenopausal females. Benign metastasizing leiomyoma (BML), first described by Steiner in 1939, is the rare finding of these smooth muscle tumors outside the uterus. It usually manifests as a single or multiple well-circumscribed nodules in young females, often occurring after a hysterectomy for a leiomyoma. It has also been observed to occur in postmenopausal women.

Patients are often asymptomatic and the tumor is found incidentally on routine imaging. However, in cases of extensive disease, patients may present with symptoms of a nonproductive cough, shortness of breath, and chest tightness.

The lungs are the most commonly reported extrauterine sites of metastases, but there are also reports of these tumors metastasizing to the mediastinum, nervous system, bones, lymph nodes, soft tissue, and the heart. Single or multiple nodules of various sizes are routinely described, but findings of a miliary pattern, cavitary nodules, and fluid containing cystic lesions have also been reported. Mediastinal adenopathy, however, is unusual.

The pathogenesis of BML remains unclear. In situ proliferation of the smooth muscles within the lungs, low-grade metastases from a leiomyosarcoma, and lymphatic spread have all been proposed. However, mechanical displacement of preexisting benign uterine tumors resulting in hematogenous spread to the lungs during hysterectomy is believed to be the most likely etiology. This is supported by the fact that these tumors develop within years (an average 15 years) of a hysterectomy and rarely after cesarean section.

VATS biopsy should be considered over transbronchial biopsy for higher tissue yield. Immunohistochemical staining is positive for actin, desmin, and vimten. Estrogen and progesterone receptors are also identified indicating a uterine source. There is sparing of the blood vessels and lymphatics in BML that differentiates it from lymphangioleiomyomatosis.

Patients with BML have an indolent course and a favorable prognosis. In some cases, spontaneous tumor regression during pregnancy and after menopause has been observed. The presence of progesterone and estrogen receptors has led to treatment options using gonadotropin-releasing hormone analogs, selective estrogen receptor modulators, progesterone therapy, and aromatase inhibitors with various success. In other cases, surgical resection of amenable lung lesions and/or oophorectomy may be beneficial.

TAKEAWAY POINTS

- Suspect BML when a young female with a history of hysterectomy for uterine leiomyoma presents with asymptomatic well-circumscribed pulmonary nodules.
- These lesions are positive for actin, desmin, and progesterone and estrogen receptors. Treatment in symptomatic patients involves surgical resection or hormonal manipulation.

FURTHER READING

Miller, J., Shoni, M., Siegert, C., Lebenthal, A., Godleski, J., McNamee, C., 2015. Benign metastasizing leiomyomas to the lungs: an institutional case series and a review of the recent literature. Ann. Thorac. Surg. 253–258.

Steiner, P.E., 1939. Metastasizing fibroleiomyoma of the uterus: Report of a case and review of the literature. Am. J. Pathol. 15 (1), 89–1107.

Case 13

A 42-year-old male is evaluated in clinic for an abnormal chest computed tomography (CT) scan after recently being admitted and managed for a right-sided pneumothorax. He currently does not complain of any pulmonary symptoms. His medical history is relevant for a history of a left-sided pneumothorax the year prior. He is a nonsmoker and does not partake in recreational drug use. He mentions that his brother also has a history of pneumothoraces.

On physical examination, there is evidence of dome-shaped, flesh-colored papules around his ears, down his neck, and upper chest. The rest of the examination is otherwise unremarkable. You review his most recent chest CT scan (Fig. 13.1).

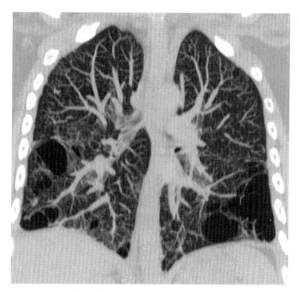

FIGURE 13.1 Coronal chest computed tomography image showing bilateral lower lobe thin-walled cystic lesions.

What is the most likely clinical diagnosis?
What test can confirm the diagnosis?

BIRT-HOGG-DUBÉ SYNDROME

The skin findings along with lower lobe cystic lesions and family history of pneumothoraces suggest the diagnosis of Birt-Hogg-Dubé (BHD) syndrome. This autosomal dominant genetic disorder was first described by three Canadian physicians in 1977. It is also known as Hornstein-Knickenberg syndrome, and fibrofolliculomas with trichodiscomas and acrochordons.

BHD is characterized by the findings of skin lesions, cystic lung disease, and renal tumors. Over 80% of patients will have multiple benign skin tumors of the hair follicles (fibrofolliculomas) that occur mainly on the scalp, face, neck, and upper chest. They are described as smooth, flesh-colored, dome-shaped growths. These skin findings are similar to the skin findings in patients with tuberous sclerosis complex syndrome. There may also be angiofibromas, or skin tags.

The majority of patients will have asymptomatic pulmonary cysts; however, spontaneous pneumothorax may occur in 24% of affected individuals. The cystic lesions in BHD are multiple, irregularly shaped, and of varying size with predominantly medial and lower lung zone location. The lesions may about the lower pulmonary arteries and veins. These characteristics may help differentiate the cysts related to BHD syndrome from other cystic lung diseases such as Langerhans cell histiocytosis and lymphangioleiomyomatosis.

Patients with BHD are at increased risk of renal tumors, mainly hybrid renal tumors consisting of chromophobe and oncoytoma cell types, as well as renal cysts. Not all patients develop renal findings. Other described manifestations of the disease include thyroid nodules, parathyroid adenomas, oral plaques, and lipomas.

Inheritance follows an autosomal dominant trait that involves mutations in the tumor suppressor gene FLCN on chromosome 17p11.2 that codes for the folliculin protein. One gene copy inheritance will result in the skin and pulmonary manifestations; however, two mutated copies of the gene are present in patients with renal tumors. Genetic testing for the FLCN gene in suspected cases confirms the diagnosis.

No specific treatments are currently available. Management is directed toward treating the different manifestations of the syndrome. Periodic monitoring for the development of renal tumors should be undertaken in affected individuals. Family genetic counseling should also be offered.

TAKEAWAY POINTS

- BHD is an autosomal dominant disorder (FLCN gene coding the folliculin protein) characterized by skin lesions, pulmonary cysts, and renal tumors.
- The pulmonary cysts are commonly found in the lower lobes adjacent to pulmonary arteries and veins.

FURTHER READING

Agarwal, P.P., Gross, B.H., Holloway, B.J., Seely, J., Stark, P., Kazerooni, E.A., 2011. Thoracic CT findings in Birt-Hogg-Dube syndrome. Am. J. Roentgenol. 196 (2), 349–352.

Birt, A.R., Hogg, G.R., Dube, W.J., 1977. Hereditary multiple fibrofolliculomas with trichodiscomas and acrochordons. Arch. Dermatol. 113 (12), 1674–1677.

Toro, J.R., Pautler, S.E., Stewart, L., Glenn, G.M., Weinreich, M., Toure, O., et al., 2007. Lung cysts, spontaneous pneumothorax, and genetic associations in 89 families with Birt-Hogg-Dube syndrome. Am. J. Respir. Crit. Care Med. 175 (10), 1044–1053.

Case 14

An 18-year-old male is admitted to the intensive care unit after being urgently intubated in the emergency department for hypoxic respiratory failure. Family members at the bedside report that he has been coughing and complaining of fatigue and shortness of breath for the last 3 days. They note that the patient started smoking cigarettes over the last few months and disclose that he has been caught a few times sniffing glue with friends. The patient is intubated and sedated.

On physical examination, he is hypertensive and auscultation of the chest has diffuse bilateral crackles. A chest X-ray shows bilateral diffuse opacities. A bedside bronchoscopy is performed with bronchoalveolar lavage (BAL). With serial lavage, the BAL fluid becomes progressively more bloody consistent with a diagnosis of diffuse alveolar hemorrhage (DAH). Laboratory testing reveals mild anemia, and acute kidney injury with creatinine of 2.4 mg/dL. Urinalysis shows red blood cell casts, proteinuria, and hematuria. Complement levels, anti-nuclear antibody and antineutrophil cytoplasmic antibody (ANCA) testing are all negative. Testing for anti-glomerular basement membrane (GBM) antibodies comes back positive.

What is the diagnosis?
What are other pulmonary-renal syndromes?

GOODPASTURE SYNDROME

Goodpasture syndrome, or anti-GBM disease, is an autoimmune pulmonary-renal syndrome characterized by pulmonary hemorrhage, progressive glomer-ulonephritis, and circulating anti-GBM antibodies. It was first described by Ernest Goodpasture in 1919 during an influenza epidemic.

The circulating autoantibodies target the alpha-3 subunit of type IV collagen on the basement membranes of the glomeruli and alveoli resulting in comple-ment cascade activation and cell injury.

Over 80% of patients have both pulmonary and renal manifestations. A lim-ited form of the disease, affecting only the kidneys or the lungs, occurs in approxi-mately 20% and 10% of patients, respectively. Symptoms may include weakness, malaise, fevers, cough, hemoptysis, dyspnea, respiratory failure, hypertension, hematuria, and acute kidney injury. Hemoptysis may be occasionally be absent in patients with DAH. Smokers and patients exposed to volatile solvents are more likely to develop pulmonary symptoms compared to nonsmokers.

Certain environmental triggers have been associated with the develop-ment of Goodpasture syndrome and DAH, including viral infections, tobacco smoking, cocaine inhalation, and inhalation of organic volatile compounds and hydrocarbons. These environmental triggers are believed to disrupt the alveolar–capillary interface, allowing the binding of anti-GBM antibodies. In this case, the recent onset of smoking and glue sniffing, with glue containing the aromatic hydrocarbon toluene, are the environmental culprits.

Other pulmonary-renal syndromes that are part of the differential diagnosis include:

● Granulomatosis with polyangiitis (formerly Wegeners granulomatosis)
● Microscopic polyangiitis
● Systemic lupus erythematosus
● Immunoglobulin A-mediated disease (Henoch-Schönlein purpura)
● Cryoglobulinemia
● Eosinophilic granulomatosis with polyangiitis (formerly Churg-Strauss syndrome)

The early recognition, diagnosis, and treatment of Goodpasture syndrome can prevent fatal deterioration. The presence of anti-GBM antibodies estab-lishes the diagnosis. In cases where laboratory testing is nondiagnostic, renal biopsy will reveal Immunoglobulin G and C3 liner deposits along the GBM. One-third of patients have been reported to also have circulating ANCAs (spe-cifically MPO-ANCA) along with the anti-GBM antibodies.

Initial therapy includes plasmapheresis, to rapidly remove the antibody, and immunosuppressive therapy, which may include corticosteroids, cyclophospha-mide, and rituximab. Immunosuppression is then continued to prevent relapse. Hemodialysis and referral for renal transplantation may be necessary when renal failure is progressive or evident on initial presentation.

TAKEAWAY POINTS

- Goodpasture syndrome is a pulmonary-renal autoimmune syndrome with anti-GBM antibodies.
- Smokers and patients with exposure to inhaled hydrocarbons are more likely to present with DAH.

FURTHER READING

Greco, A., Rizzo, M.I., De Virgilio, A., Gallo, A., Fusconi, M., Pagliuca, G., et al., 2015. Goodpasture's syndrome: a clinical update. Autoimmun. Rev. 14 (3), 246–253.

McCabe, C., Jones, Q., Nikolopoulou, A., Wathen, C., Luqmani, R., 2011. Pulmonary-renal syndromes: an update for respiratory physicians. Respir. Med. 105 (10), 1413–1421.

Case 15

A 26-year-old female is seen in clinic for exertional shortness of breath. She has experienced shortness of breath with activity since childhood. Her symptoms have slowly progressed over the last few years; she has noticed early onset of symptoms and occasional wheezing with exercise. She was diagnosed with asthma during adolescence and is on bronchodilator therapy. She had a tonsillectomy at the age of 18 and was informed that the anesthesiologist had difficulty inserting a size 6.5 endotracheal tube.

On examination, vital signs are within normal range and the patient is not in any distress. Auscultation of the chest reveals normal breath sounds but an inspiratory and expiratory stridor is heard over the neck. Pulmonary function testing shows a mild obstructive ventilatory defect and flattening of the inspiratory and expiratory limbs of the flow-volume loop. Diffusing capacity was normal. A chest X-ray was unrevealing. The patient underwent bronchoscopy that revealed narrowing of the trachea with the presence of complete tracheal rings extending half way down her trachea.

What is the diagnosis?
What are some other congenital tracheal anomalies?

COMPLETE TRACHEAL RINGS

The normal trachea consists of C-shaped cartilages with a membranous posterior portion. Complete or near-complete tracheal rings are rare tracheal malformations characterized by the absence of this posterior membranous portion resulting in complete circular tracheal cartilage. This may involve a portion or the entire trachea (known as stove-pipe trachea).

This finding is occasionally associated with other congenital cardiac and pulmonary artery anomalies, with half of cases having an aberrant left pulmonary artery. These cases are commonly present early in childhood but a few cases have gone unrecognized till early adulthood, being mislabeled as asthma.

Symptoms include dyspnea on exertion and stridor that are exacerbated with physical activity or with respiratory infections. As in our case, patients may have a history of being difficult to intubate. Difficult intubation may also be a clue to a proximal tracheal stenotic lesion.

Diagnosis is suggested by visualization of complete tracheal cartilages on chest computed tomography scan imaging and is confirmed by bronchoscopy. Surgical tracheoplasty of the stenotic tracheal segments may be warranted to alleviate symptoms if they are severe enough to interfere with the activities of daily living. Conservative management is reserved for asymptomatic patients and those with minimal symptoms.

Other congenital tracheal anomalies:

- Idiopathic tracheal stenosis
- Congenital tracheobronchomalacia
- Tracheal bronchus (a bronchus arising from the distal trachea)
- Tracheal diverticulum
- Tracheal cartilaginous sleeve: vertical fusion of the tracheal cartilage results in a smooth narrow trachea without rings or ridges. Without intervention, this malformation is incompatible with life.

TAKEAWAY POINTS

- Complete tracheal rings are rare congenital abnormality that may be misdiagnosed as asthma. A history of resistance during intubation may provide a clue toward making the diagnosis.
- It is commonly associated with cardiac and pulmonary vascular anomalies. Tracheoplasty can be undertaken to alleviate symptoms.

FURTHER READING

Aneeshkumar, M.K., Ghosh, S., Osman, E.Z., Clarke, R.W., 2005. Complete tracheal rings: lower airway symptoms can delay diagnosis. Eur. Arch. Otorhinolaryngol. 262 (2), 161–162.

Mehta, A.C., Thaniyavarn, T., Ghobrial, M., Khemasuwan, D., 2015. Common congenital anomalies of the central airways in adults. Chest 148 (1), 274–287.

Rutter, M.J., Cotton, R.T., Azizkhan, R.G., Manning, P.B., 2003. Slide tracheoplasty for the management of complete tracheal rings. J. Pediatr. Surg. 38 (6), 928–934.

Case 16

A 41-year-old female with history of recurrent upper respiratory tract infections is referred for evaluation of a persistent nonproductive cough for the last 4 months and an abnormal chest computed tomography (CT) scan (Fig. 16.1). She also experiences shortness of breath after climbing two flights of stairs but denies chest pain, fevers, chills, night sweats, or hemoptysis. She has no history of tobacco or illicit drug use, no travel in the last year. Her family history is noncontributory.

Pulmonary function tests show a restrictive ventilator defect with a mild reduction in diffusing capacity of the lungs for carbon monoxide (DLCO). Workup including sputum cultures for bacteria and acid-fast bacteria, autoimmune serology, alpha-1 antitrypsin phenotype, and cystic fibrosis testing are all negative. All her immunoglobulin levels are found to be low and a diagnosis of common variable immunodeficiency (CVID) is made. The patient undergoes bronchoscopy with transbronchial biopsies, which reveal evidence of lymphocytic bronchiolitis, interstitial and organizing pneumonia, and poorly formed nonnecrotizing granulomas (Figs. 16.2 and 16.3). No organisms are cultures from the bronchoalveolar lavage fluid and biopsy samples.

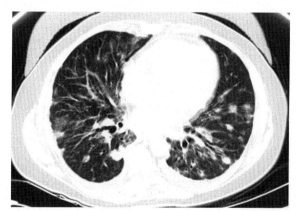

FIGURE 16.1 Chest computed tomography scan with scattered nodules, ground-glass infiltrates, and bronchiectasis, all favoring the lower lobes bilaterally.

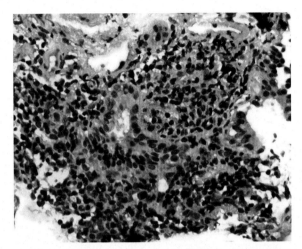

FIGURE 16.2   Biopsy revealing evidence of lymphocytic bronchiolitis.

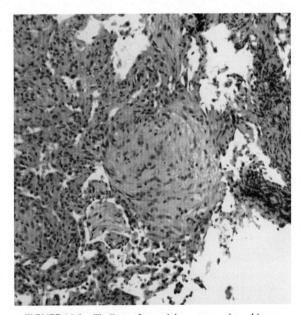

FIGURE 16.3   Findings of organizing pneumonia on biopsy.

What is the diagnosis?

GRANULOMATOUS-LYMPHOCYTIC INTERSTITIAL LUNG DISEASE

CVID is the most common form of primary immunodeficiency characterized by hypogammaglobulinemia, as a result of B cell differentiation failure, along with T lymphocyte abnormalities. Patients with CVID are predisposed to recurrent upper and lower respiratory tract infections and bronchiectasis, but noninfectious interstitial lung disease may develop in 10–30% of cases.

The presence of noninfectious interstitial lung disease in patients with CVID with a histological pattern of lymphocytic interstitial pneumonia (LIP), follicular bronchiolitis, and lymphoid hyperplasia has been termed granulomatous-lymphocytic interstitial lung disease (GLILD). The etiology of GLILD is not known, although it has been suggested that viral infections such as EBV, HIV, and HHV-8 may play a role in its pathogenesis. The occurrence of GLILD in other forms of immunodeficiencies has also been reported.

The most common symptoms of GLILD are shortness of breath and chronic cough. Pulmonary function tests may reveal a restrictive ventilatory defect with a reduced diffusing capacity (DLCO), but are normal in some cases. CT findings vary widely with lower lung predominant pulmonary nodules being the most commonly reported finding. Interlobular septal thickening, ground-glass opacities, bronchiectasis, multifocal pulmonary consolidation, and mediastinal lymphadenopathy may also be present. Lower lung zones are preferentially involved, in contrast to sarcoidosis.

On histological analysis, there is evidence of lymphocytic infiltration, with a nodular peribronchiolar and interstitial distribution being the predominant pattern of lymphoid proliferation. Poorly formed nonnecrotizing granulomas with sporadic giant cells, organizing pneumonia, and interstitial fibrosis may also be present.

The presence of ILD in patients with CVID portends a worse prognosis. No treatment guidelines are currently established for treatment of GLILD. Corticosteroids are usually given first. In more advanced lung disease, the use of combined therapy with other immunosuppressive agents such as cyclosporine, azathioprine, methotrexate, and rituximab has been reported. Serial monitoring of lung findings is indicated since these patients are at higher risk of developing lymphoma.

Our patient was started on intravenous immunoglobulin therapy along with corticosteroids with significant improvement in her symptoms.

TAKEAWAY POINTS

- CVID is the most common primary immunodeficiency, and a small number of patients with CVID develop interstitial lung disease termed GLILD.
- GLILD preferentially involves the lower lung zones and histologically has findings of LIP, follicular bronchiolitis, and lymphoid hyperplasia.

FURTHER READING

Boursiquot, J.N., Gerard, L., Malphettes, M., Fieschi, C., Galicier, L., Boutboul, D., et al., 2013. Granulomatous disease in CVID: retrospective analysis of clinical characteristics and treatment efficacy in a cohort of 59 patients. J. Clin. Immunol. 33 (1), 84–95.

Park, J.H., Levinson, A.I., 2010. Granulomatous-lymphocytic interstitial lung disease (GLILD) in common variable immunodeficiency (CVID). Clin. Immunol. 134 (2), 97–103.

Case 17

A 58-year-old female presents complaining of a 10-month history of worsening weakness, shortness of breath on exertion, and nonproductive cough. She first noticed generalized weakness, difficulty climbing stairs, and getting up from a chair. Over time she began having progressive difficulty breathing with activity. She also notes arthritis in the small joints of her hands. She is a nonsmoker and has no relevant medical or surgical history, no family history of lung or autoimmune disease, and no recent travel.

On physical examination, there are bilateral basal fine inspiratory crackles on auscultation of the lungs. Heart sounds are normal with no abnormal sounds. Thickened, fissured skin is noted on the radial aspect of her fingertips, thumbs, and palms. No other skin findings are present. She has great difficulty standing up from a chair and requires the support of her hands.

Pulmonary function tests (PFTs) show a moderate restrictive ventilator defect. A chest X-ray reveals a bilateral basal interstitial lung disease process. High-resolution computed tomography (HRCT) scan of the chest is ordered and shown in Fig. 17.1. Her creatine kinase and aldolase levels are elevated but testing for antinuclear antibody, anti-dsDNA, anti-centromere, anti-Scl-70, anti-U1-RNP, anti-Ro, and anti-La are all negative.

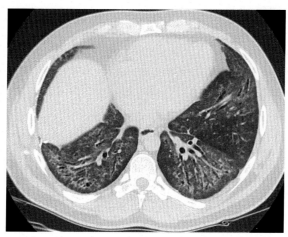

FIGURE 17.1   High-resolution computed tomography image of the chest showing bilateral basal ground-glass opacities, reticulations, and traction bronchiectasis. There is sparing of the immediate subpleural region. These findings are consistent with nonspecific interstitial pneumonia.

What is the most likely diagnosis?
What additional serologic testing should be performed?

Rare and Interesting Cases in Pulmonary Medicine.

ANTISYNTHETASE SYNDROME

Antisynthetase syndrome is an autoimmune disorder considered to be a subgroup of the idiopathic inflammatory myopathies that also includes polymyositis and dermatomyositis.

The antisynthetase syndrome is more common in women and is characterized by the constellation of myositis, interstitial lung disease, polyarthritis, fevers, mechanic's hands, and Raynaud phenomenon. Generally, myositis precedes or parallels lung involvement, and significant variability in the disease course and severity exists.

Antibodies to the anti-aminoacyl-transfer RNA (tRNA) synthetases are present in the serum. The most commonly detected antibody is the anti-Jo-1 (anti-histidyl-tRNA synthetase), named in 1980 after the first patient, John P, in whom the antibody was detected. Other antisynthetase antibodies include anti-PL-7, anti-PL-12, anti-OJ, anti-EJ, anti-KS, anti-ZO, anti-YRA, and anti-Wa antibodies. Patients with either anti-PL-7 or anti-PL-12 antibodies may have interstitial lung disease without muscle involvement (amyopathic interstitial lung disease). Anti-Jo-1 antibody is detected in up to 70% of patients with myositis with concomitant interstitial lung disease, while the other antibodies are present in only 3% of patients with myositis.

Clinically, proximal muscle weakness, predominately of the iliopsoas and quadriceps muscles, may manifest as difficulty climbing stairs or standing up unsupported from a sitting position. Serum creatine phosphokinase and aldolase are often elevated in the setting of active myositis. Patients may have skin manifestations of Raynaud phenomenon, Gottron papules, heliotrope rash, and mechanic's hand. Mechanic's hands develop in 30% of patients and manifests as thickened, fissured, and hyperkeratotic skin at the tips and margins of the fingers, particularly at the radial aspect of the index fingers and palms. A nonerosive symmetrical polyarthritis, mostly affecting the small joints, occurs in approximately half of patients.

Most patients develop interstitial lung disease, which is associated with significant morbidity. Patients usually present with insidious symptoms of shortness of breath on exertion and a nonproductive cough. PFTs will reveal a restrictive lung process while on HRCT a pattern of nonspecific interstitial pneumonia (NSIP) with bilateral, basal predominant ground-glass opacification is most typical. A pattern of usual interstitial pneumonitis or organizing pneumonia may also be seen. Patients may also develop pulmonary arterial hypertension regardless of lung involvement.

Patients often respond to glucocorticoid therapy with improved muscle strength and respiratory status, but prolonged therapy is usually required. Additional immunosuppressive therapy may be required for disease control in advanced cases. The use of azathioprine, methotrexate, cyclophosphamide, or rituximab has been reported. If pulmonary disease is progressive, referral to a lung transplantation center may be considered.

TAKEAWAY POINTS

- Antisynthetase syndrome should be considered in the differential diagnosis of patients with interstitial lung disease and proximal myositis. Anti-Jo-1 is the most commonly associated antibody with this disorder.
- Amyopathic interstitial lung disease can occur in association with anti-PL-7 and anti-PL-12 antibodies.
- In patients with interstitial lung disease, an NSIP pattern is the most frequent on HRCT.

FURTHER READING

Katzap, E., Barilla-LaBarca, M.L., Marder, G., 2011. Antisynthetase syndrome. Curr. Rheumatol. Rep. 13 (3), 175–181.

Nishikai, M., Reichlin, M., 1980. Heterogeneity of precipitating antibodies in polymyositis and dermatomyositis. Characterization of the Jo-1 antibody system. Arthritis Rheum. 23 (8), 881–888.

Case 18

A 59-year-old male is evaluated for shortness of breath after undergoing biliary stent placement for a stricture in the biliary system. A chest X-ray is done and shows a right pleural effusion that was not present prior to his procedure. The patient is afebrile and has no signs or symptoms of an underlying infection. He has no abdominal pain, but experiences shortness of breath on minimal exertion, associated with a drop in his oxygen saturation.

A thoracentesis is performed that reveals straw-colored fluid that is an exudate by Light's criteria. The pleural bilirubin level is found to be elevated at 23 mg/dL with a serum bilirubin level of 3.6 mg/dL.

What type of pleural effusion does the patient have?
How would you manage his pleural effusion?

BILOTHORAX

Bilothorax, also known as bilious pleural effusion, thoracobilia, or cholethorax, is a rare type of pleural effusion characterized by the presence of bile in the pleural cavity. It was first described by Rowe in 1989, in a patient who developed bilateral pleural effusions after a gastrectomy.

Bilothorax should be suspected in the setting of development of a new pleural effusion after a hepatobiliary procedure or abdominal surgery. It has been reported to occur as a complication of hepatic trauma, hepatic tumor or abscess rupture, intraabdominal surgery, and percutaneous transhepatic biliary drainage. Bile leakage into the pleural space arises as a result of damage to the pleural cavity, due to formation of a biliary-pleural or cholecystopleural fistula. Bile flow may also occur through a congenital or acquired diaphragmatic communication, resulting in the formation of a bilothorax. The pleural effusion is almost always right sided but may also occur on the left or bilaterally. Diagnosis is established by an elevated pleural bilirubin level with a pleural bilirubin to serum bilirubin ratio >1. In cases where the diagnosis remains questionable, a hepatobiliary scintigraphy scan can reveal bile leakage into the thoracic space and may help determine fistula location to assist in surgical correction.

Patients will usually complain of shortness of breath or chest heaviness and may develop respiratory failure. Abdominal pain may occur in the setting of bilious ascites. The presence of fever should raise suspicion for an infected pleural space. On drainage, the pleural fluid may appear straw colored or in some cases green. In cases of a cholecystopleural fistula, gallstones may be recovered from the pleural space.

First-line treatment involves chest tube placement to drain the pleural space followed by treatment of the underlying etiology of the bilothorax. As the bile is often infected, bacterial infection occurs in half of patients with bilothorax and antibiotic therapy is indicated. In such cases, the pleural space should be treated as an empyema. Bilothorax has been reported to occur with associated biliary ascites; in such cases, the ascites should be drained as well.

TAKEAWAY POINTS

- Bilothorax is the presence of bile in the pleural space and should be suspected if a new pleural effusion develops after a biliary procedure or abdominal surgery or in the context of biliary peritonitis.
- A pleural bilirubin to serum bilirubin ratio greater than 1 confirms the diagnosis.
- Management involves chest drainage of the effusion, treating the underlying cause, and antibiotics if the patient has symptoms of infection.

FURTHER READING

Delco, F., Domenighetti, G., Kauzlaric, D., Donati, D., Mombelli, G., 1994. Spontaneous biliothorax (thoracobilia) following cholecystopleural fistula presenting as an acute respiratory insufficiency. Successful removal of gallstones from the pleural space. Chest 106 (3), 961–963.

Rowe, P.H., 1989. Bilothorax–an unusual problem. J. R. Soc. Med. 82 (11), 687–688.

Case 19

A 38-year-old male is referred for evaluation of incidentally found mediastinal adenopathy on computed tomography (CT) imaging done after a motor vehicle accident. He has no relevant past medical history. He works as an accountant, does not smoke, has no significant exposures, and no family history of lung disease or cancer. He has no complaints of fevers, chills, night sweats, weight loss, chest pain or discomfort, shortness of breath, coughing, or wheezing. On physical examination, no palpable lymphadenopathy or hepatosplenomegaly is noted and the remainder of the examination is normal.

The CT scan of the chest shows mediastinal adenopathy but no parenchymal abnormalities. There is no evidence of enlarged lymph nodes elsewhere in the body. Due to the concern for possible malignancy, an endobronchial ultrasound-guided transbronchial needle aspiration of the mediastinal adenopathy is performed. Pathology showed no evidence of malignancy or lymphoma; however, there was benign enlargement of the lymph node with increased number of follicles and hyaline material and blood vessels, consistent with Castleman disease.

What are the different types of Castleman disease?
What viruses are other syndromes are associated with this disease?

CASTLEMAN DISEASE

Castleman disease, also known as angiofollicular lymph node hyperplasia, is a benign lymphoproliferative disorder that was first described by Benjamin Castleman in 1954.

The disease occurs as a result of hyperactivation of the immune system secondary to the overproduction of cytokines, mainly vascular endothelial growth factor and interleukin-6 (IL-6). The disease is either localized to one part of the body, termed unicentric Castleman disease (UCD), or is widespread affecting different regions, termed multicentric Castleman disease (MCD). The most common locations of lymph node proliferation are in the thorax, neck, and abdomen.

UCD variant is the most common type of the disease. Most patients with UCD are asymptomatic, but some may have symptoms secondary to lymph node enlargement causing compressive or obstructive symptoms. Follicular dendritic cell sarcoma, a rare malignancy of the lymph nodes, and paraneoplastic pemphigus, a mucocutaneous blistering autoimmune skin disease, both have been associated with UCD.

MCD has been linked with HIV and HHV-8 infection. Patients present with constitutional symptoms such as fevers, weight loss, and night sweats. MCD has been associated with hemolytic anemia, hypergammaglobulinemia, and hepatosplenomegaly. Patients with MCD are further classified according to their HIV and HHV-8 status (e.g., HHV-8 positive MCD or HHV-8 negative MCD). MCD has also been associated with polyneuropathy, organomegaly, endocrinopathy, monoclonal gammopathy, and skin changes syndrome, and thrombocytopenia, anasarca, fevers, reticulin fibrosis, and organomegaly syndrome also known as Castleman-Kojima disease. Both variants of the disease have been associated with an increased risk of malignancy, specifically Kaposi sarcoma and non-Hodgkin lymphoma.

CT findings will show mediastinal and hilar adenopathy, and in some cases parenchymal findings of interlobular septal thickening, ground-glass opacities, subpleural nodules, areas of consolidation, and bilateral pleural effusions. In MCD, the lymph nodes are positron emission tomography-avid with a standardized uptake value in the range of 2.5–5.

Histologically, there are four subtypes:

- Hyaline vascular type constitutes 90% of cases. It commonly occurs in a localized form and portends a good prognosis.
- Plasma cell type is often multicentric with patients usually symptomatic.
- Mixed subtype is a rare form with both hyaline vascular and plasma cell types.
- Plasmablastic type is a symptomatic multicentric variant with poor prognosis.

The serum IL-6 levels and HHV-8 viral load may be elevated. Definitive diagnosis is made on lymph node excision or biopsy. There is polyclonal nodal expansion leaving the structure of the lymph node generally intact with regression of the germinal center (Fig. 19.1). Histological features of either hyaline vascular or plasma cell variant may be present. HHV-8 staining in the lymph node may further assist in diagnosis.

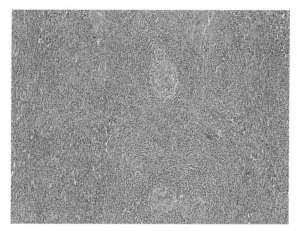

FIGURE 19.1   Lymph node with regressed follicles with atretic germinal center showing expanded mantle zone with concentric rings of small lymphocytes. *Courtesy of Humberto Trejo Bittar, MD.*

Treatment of UCD involves surgical resection with excellent prognosis. In cases of incomplete resection or remission, radiation therapy can be utilized. For MCD, because of the rarity of the disease, there is not a proven treatment. Patients should be encouraged to enroll in a clinical trial. Glucocorticoids, immunotherapy alone or in combination with antiviral drugs and/or chemotherapy have been used depending upon the viral status, performance status, and extent of disease. There has been success with the use of IL-6 directed therapy such as siltuximab and tocilizumab in treating MCD.

TAKEAWAY POINTS

- Castleman disease is a lymphoproliferative disorder with two variants, unicentric or multicentric disease, and four distinct histological subtypes.
- Both variants of the disease have been associated with different syndromes. Multicentric disease has been linked to HIV and HHV-8 infection.
- Unicentric disease has a better prognosis with surgical treatment as an option. Multicentric disease portends a worse prognosis and is treated with glucocorticoids, immunotherapy, chemotherapy, antiviral therapy, and combinations.

FURTHER READING

Andhavarapu, S., Jiang, L., 2013. POEMS syndrome and Castleman disease. Blood 122 (2), 159.

Castleman, B., Towne, V., 1954. CASE records of the Massachusetts General Hospital Weekly Clinicopathological Exercises: Case 40011. N. Engl. J. Med. 250 (1), 26–30.

Kawashima, M., Usui, T., Okada, H., Mori, I., Yamauchi, M., Ikeda, T., et al., 2015. TAFRO syndrome: 2 cases and review of the literature. Mod. Rheumatol. 1–5.

Soumerai, J.D., Sohani, A.R., Abramson, J.S., 2014. Diagnosis and management of Castleman disease. Cancer Control 21 (4), 266–278.

Case 20

A 19-year-old male is seen in hospital for evaluation of a right-sided pleural effusion. He initially presented with symptoms of a nonproductive cough and shortness of breath over a span of 5 days. He also complained of right shoulder and right chest pain for the last month. He has no relevant medical history of note. Physical examination showed dullness and diminished breath sounds on the right but no other abnormal findings.

A chest X-ray showed a right pleural effusion of moderate size and osteolytic lesions of the right humeral head and right rib that are correlated with the locations of the patient's pain. A computed tomography scan of the chest confirmed the right pleural effusion and the osteolytic lesions, but did not reveal any parenchymal abnormality, thoracic adenopathy, or pulmonary embolism. A thoracentesis was performed, and the pleural fluid was milky white with elevated triglyceride levels, consistent with a chylothorax. A thorough work-up for a malignancy did not reveal any masses or lesions elsewhere in the body. The patient underwent a bone biopsy of the right humerus lesion that revealed increased vascularization of the bone with proliferation of lymphatic vessels and diffuse hemangiomatosis.

What is the diagnosis?

GORHAM-STOUT SYNDROME

Gorham-Stout syndrome is a rare disease characterized by the nonmalignant proliferation of vascular and lymphatic elements in the bone resulting in their spontaneous resorption. As such, it is also termed vanishing bone syndrome, phantom bone disease, idiopathic massive osteolysis, or Gorham massive osteolysis. The first description of the syndrome is believed to be by Jackson in 1838, but the name was later coined after Gorham and Stout reported two cases in 1955.

The etiology of this syndrome remains unknown but association with PTEN germ line mutation has been suggested. It may manifest in childhood and early adulthood and affects males and females equally. There is proliferation of lymphatics and blood vessels that invade the bone and surrounding tissue. Circumscribed spontaneous bone resorption occurs, most commonly in the long bones and bones of the skull, mandible, pelvis, and shoulder.

Patients will often complain of pain at the site of bone involvement. Skeletal deformities and pathological fractures may also occur. Chest involvement, as a result of thoracic duct invasion or extension of lymphatic dysplasia into the pleural cavity, results in chylothorax formation, reported to occur in 17% of cases. Thoracic involvement is associated with symptoms of coughing, dyspnea, and respiratory failure and portends a poor prognosis. All reported patients with chylothorax had bone involvement of either their ribs, scapula, clavicle, or thoracic vertebrae. Cases of chylopericardium have also been reported in the literature.

Diagnosis can be confirmed on bone biopsy of affected regions showing areas of angiomatosis. In later stages, the osseous tissue is replaced by fibrous tissue. Findings of lymphangiomatosis and hemangiomatosis may also be present on biopsy.

All patients with pleural effusion should have a thoracentesis to determine the presence of chylothorax. If present, further evaluation of the chylothorax with a lymphangiogram study should be undertaken to determine lymphatic anatomy to assist in surgical correction. Early thoracic duct ligation should be considered in patients with chylothorax. Other treatment options include dietary fat restriction with medium-chain triglycerides supplementation, pleurodesis, pleurectomy, and somatostatin therapy. Specific treatment for the bone involvement includes surgical resection and/or the use of bisphosphonates, radiation treatment, and interferon alfa-2b. There is one case report of successful treatment with bevacizumab. Cases of spontaneous recovery of part of the lost bone have also been reported.

TAKEAWAY POINTS

- Gorham-Stout syndrome is a disorder of unclear etiology characterized by massive bone osteolysis.
- Chylothorax occurs in 17% of patients, usually with thoracic bone involvement, and is associated with a poor prognosis.

FURTHER READING

Jackson, J.B.S., 1838. A boneless arm. Boston Med. Surg. J. 18, 368–369.

Gorham, L.W., Stout, A.P., 1955. Massive osteolysis (acute spontaneous absorption of bone, phantom bone, disappearing bone); its relation to hemangiomatosis. J. Bone Joint Surg. Am. 37-a (5), 985–1004.

Tie, M.L., Poland, G.A., Rosenow 3rd, E.C., 1994. Chylothorax in Gorham's syndrome. A common complication of a rare disease. Chest 105 (1), 208–213.

Case 21

A 55-year-old male is admitted with fatigue and altered mental status. His family states that he has been having a persistent nonproductive cough for over a year. Initial blood work reveals that the patient is hypoglycemic (glucose 36 mg/dL). He promptly responds and is more alert shortly after administering an ampule of D50. A chest X-ray shows a mass in the right lung. A computed tomography (CT) scan of the chest confirms a pleural-based right lung mass without parenchymal involvement (Fig. 21.1). The patient has no previous imaging for comparison. A biopsy is done showing variable spindle-shaped cells that stain positive for CD34 and vimentin but not for cytokeratin.

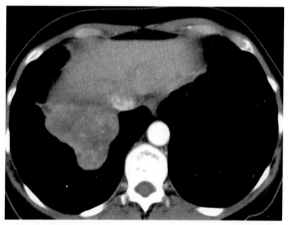

FIGURE 21.1   A right-sided heterogeneous pleural-based mass that is nearly indistinguishable from the liver. *Courtesy of Tan-Lucien Mohammed, MD.*

What is the diagnosis?
What is the described paraneoplastic syndrome associated with this tumor?

SOLITARY FIBROUS TUMOR OF THE PLEURA

Solitary fibrous tumor of the pleura (SFTP), also known as benign fibrous tumor of the pleura or pleural fibroma, is a benign, slow-growing tumor arising from the mesenchymal cells of the pleura that was first described in 1931. It accounts for <5% of all pleural tumors and is not associated with smoking or asbestos exposure. Approximately 10–30% of cases are reported to be malignant.

These tumors occur more commonly in females, and most patients are diagnosed in their 50s to 70s. Almost 80% of cases arise from the visceral pleura with a predilection to involve the middle and lower lung zones.

Over half of patients are asymptomatic, with the tumor being incidentally found on imaging of the chest. In some cases, patients may be symptomatic, complaining of chest pressure or pain, a nonproductive cough, or shortness of breath. Chest pain is more common when the tumor arises from the parietal pleura.

Associated paraneoplastic syndromes include:

- Doege–Potter syndrome, severe symptomatic hypoglycemia related to excess production of insulin-like growth factor II, occurs in 3–5% of cases.
- Digital clubbing and hypertrophic pulmonary osteopathy (HPOA; Pierre-Marie-Bamberger syndrome) occurs in 22% of patients.

CT scan of the chest is the gold standard imaging test and it also assists in guiding surgical resection. It will reveal pleural-based or pedunculated, well-circumscribed or lobulated mass with heterogeneous attenuation. In some instances, the tumor may arise within a fissure and appear to be invading the lung parenchyma, termed "inverted tumor." Up to two-thirds of tumors may enhance if contrast is administered.

Histology reveals irregularly arranged spindle-shaped cells on a collagenous background (Fig. 21.2). Immunohistochemical staining is positive for CD34, bcl-2, and vimetin expression. Markers for cytokeratin, S100, actin, and desmin are negative. It has been reported that nuclear expression of C-terminal of STAT6 is a very specific and sensitive marker for SFTP. These markers can help differentiate SFTP from mesothelioma and other tumors.

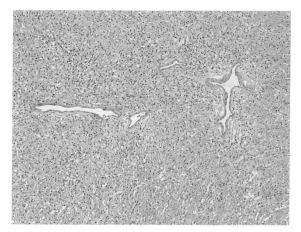

FIGURE 21.2   Proliferation of oval spindle cells not arranged in any pattern with branching "staghorn" vessels. There is no necrosis or significant mitotic activity. *Courtesy of Humberto Trejo Bittar, MD.*

The prognosis of SFTP is good with a 10-year survival rate of up to 98%. Treatment of choice is surgical resection via a wedge resection, with a high cure rate. However, in 8% of surgically treated cases, the tumor may recur. The associated paraneoplastic syndromes, if present, resolve after tumor resection.

TAKEAWAY POINTS

- SFTP is a benign pleural-based tumor that can be differentiated from other tumors based on immunohistochemistry: positive for CD34, bcl-2, vimentin, and STAT6, but negative for cytokeratin.
- Associated uncommon paraneoplastic syndromes include severe hypoglycemia and HPOA. These usually resolve with tumor resection.

FURTHER READING

Cardillo, G., Carbone, L., Carleo, F., Masala, N., Graziano, P., et al., 2009. Solitary fibrous tumors of the pleura: an analysis of 110 patients treated in a single institution. Ann. Thorac. Surg. 88 (5), 1632–1637.

Klemperer, P., Rabin, C.B., 1931. Primary neoplasms of the pleura. A report of five cases. Arch. Pathol. 11, 385–412.

Case 22

A 33-year-old male presents with shortness of breath on exertion for 3-week duration. He has no relevant medical or family history, is a smoker with a 10 pack-year smoking history, and no history of illicit drug use. The patient is hemodynamically stable. Physical examination reveals hyperresonance on percussion of the chest bilaterally as well as diminished breath sounds in the upper lung fields. There is no tracheal shift on examination. A chest X-ray and chest computed tomography scan are shown in Figs. 22.1 and 22.2.

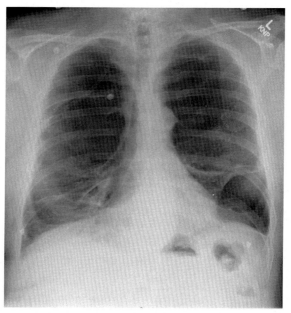

FIGURE 22.1   Chest X-ray revealing extensive bilateral upper lobe bullous disease of the lungs.

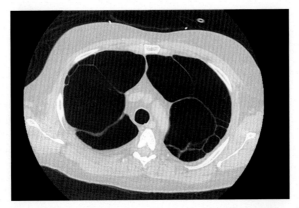

FIGURE 22.2   Chest computed tomography scan showing bilateral bullous disease.

What is the diagnosis?

VANISHING LUNG SYNDROME

Vanishing lung syndrome (VLS), type I bullous disease, or idiopathic giant bullous emphysema, is an uncommon disease characterized by giant bullae that compress the adjacent lung parenchyma giving an appearance of a disappearing lung parenchyma that may be mistaken for a pneumothorax. It was first described by Burke in 1937 in a young male with bullae occupying two-thirds of his hemithorax.

VLS usually occurs in young, thin male smokers, in contrast to smoking-related emphysema that occurs in older populations. VLS may also occur in nonsmokers and has been associated with connective tissue disorders such as Marfan and Ehler-Danlos syndromes, and with alpha-1 antitrypsin deficiency. Affected individuals may be asymptomatic or may complain of shortness of breath, cough, and chest pain.

The giant bullae (defined as occupying over at least one-third of the hemithorax) usually occur as a result of paraseptal emphysematous blebs merging to form the giant bullae. The upper lobes are most often involved with compression of adjacent structures resulting in atelectasis, mediastinal shift, or an inverted diaphragm. The bullae usually present bilaterally, but in an asymmetrical pattern with one lung being affected to a larger extent than the other. Due to the large size that these bullae reach, they may be mistaken for a pneumothorax in symptomatic patients and may result in chest tube placement and further complications. As such, distinction from a pneumothorax due to the presence of a pleural line or double-wall sign on CT scan can help make the distinction.

VLS is progressive disorder and potential complications include the development of pneumothorax, infected bullae, compression of adjacent lung, and increased risk of lung cancer. Bullectomy is the surgical procedure of choice.

TAKEAWAY POINTS

- Vanishing lung syndrome is characterized by the presence of giant bullae, occupying at least one-third of the hemithorax, resulting in compression of adjacent structures.
- It is important to distinguish the giant bullae from pneumothorax to avoid inadvertent placement of a chest tube. Bullectomy is the treatment of choice.

FURTHER READING

Burke, R., 1937. Vanishing lungs: a case report of bullous emphysema. Radiology 367–371.

Palla, A., Desideri, M., Rossi, G., Bardi, G., Mazzantini, D., et al., 2005. Elective surgery for giant bullous emphysema: a 5-year clinical and functional follow-up. Chest 128, 2043–2050.

Stern, E.J., Webb, W.R., Weinacker, A., Muller, N.L., 1994. Idiopathic giant bullous emphysema (vanishing lung syndrome): imaging findings in nine patients. Am. J. Roentgenol. 162 (2), 279–282.

Sharma, N., Justaniah, A., Kanne, J.P., Gurney, J.W., Mohammed, T.-L.H., 2009. Vanishing lung syndrome (giant bullous emphysema): CT findings in 7 patients and a literature review. J. Thorac. Imaging 24, 227–230.

Case 23

A 21-year-old male with no significant past medical history presents to the hospital after a motor vehicle injury. The patient reported losing control of his car and hitting a tree at 55 mph without wearing his seatbelt. He only remembers hitting his chest on the steering wheel before being brought in by ambulance. His urine toxicology was negative and he has no significant environmental or occupational exposures. His chest computed tomography (CT) scan is shown (Fig. 23.1). The patient is complaining only of left chest pain and denies any fevers, cough, hemoptysis, or other symptoms of note.

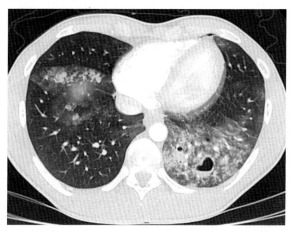

FIGURE 23.1   Chest computed tomography scan showing an area of consolidation in the left lower lobe surrounding a cavity containing an air–fluid level.

What is the etiology of the chest findings?
How should the patient be treated?

TRAUMATIC PULMONARY PSEUDOCYST

Chest trauma, whether blunt or penetrating, can result in various clinical pulmonary manifestations and complications. Traumatic pulmonary pseudocysts (TPP), also known as traumatic pneumatoceles, are an uncommon presentation after chest trauma occurring in 1–3% of cases.

TPP, first described by Fallon in 1940, usually occur after blunt chest trauma, but may rarely occur after penetrating chest injury. Primary TPPs occur due to damage and laceration of the lung parenchyma, as a result of rapid compression and decompression injury. Secondary TPPs develops after the resolution of a pulmonary hematoma.

TPP occur within a few hours of an injury, and are seen more frequently in younger individuals with a male predominance. The higher lung compliance in children and young adults seems to predispose them to the development of TPP. Patients are often asymptomatic with the TPP being detected on imaging. Symptomatic patients may complain of chest pain, cough, hemoptysis, or dyspnea.

TPPs can be seen on a chest X-ray, but often require chest CT scan for detection. Imaging shows thin-walled cavity or cavities, ranging from 2 to 14 cm in size, often containing an air–fluid level. TPP more frequently develop in the lower lobes and may occur at the site of injury or on the opposite side of chest. The finding of single or multiple cystic lesions surrounded by a consolidated lung parenchyma, within hours after chest trauma, is diagnostic for TPP. Cavitation within a hematoma is a later finding. Infectious etiologies should always be considered before making the diagnosis.

Patients with TPP require no intervention besides symptom management, since these lesions usually resolve within 6 weeks. Lesions that persist after 4 months should prompt consideration for other etiologies. Occasionally, TPP may result in complications requiring medical or surgical treatment. TPP may become infected and may lead to the development of a lung abscess, requiring medical treatment with antibiotics. Pneumothorax may also develop requiring surgical management in some patients.

TAKEAWAY POINTS

- TPP are rare benign lesions that occur after blunt chest trauma and resolve within a few weeks.
- The presence of a cysts surrounded by consolidated lung on chest CT after chest trauma is often diagnostic for TPP.

FURTHER READING

Chon, S.H., Lee, C.B., Kim, H., Chung, W.S., Kim, Y.H., 2006. Diagnosis and prognosis of traumatic pulmonary psuedocysts: a review of 12 cases. Eur. J. Cardiothorac. Surg. 29 (5), 819–823.

Fallon, M., 1949. Lung injury in intact thorax with report of case. Br. J. Surg. 28, 39–49.

Tsitouridis, I., Tsinoglou, K., Tsandiridis, C., Papastergiou, C., Bintoudi, A., 2007. Traumatic pulmonary pseudocysts: CT findings. J. Thorac. Imaging 22 (3), 247–251.

Case 24

A 64-year-old male is admitted with fevers, chills, a nonproductive cough, and shortness of breath that have been slowly progressive over 2 weeks. Sputum and blood cultures are sent and he is started on broad-spectrum antibiotics. A computed tomography (CT) scan of the chest is shown (Fig. 24.1). The patient does not show any improvement after a few days of antibiotics and undergoes a bronchoscopy with transbronchial biopsies.

Infectious work-up is unrevealing, but the biopsy results show evidence of intraalveolar fibrin deposits and areas of organizing pneumonia (Fig. 24.1). There is no evidence of hyaline membranes or eosinophils on biopsy. Based on these results, the patient is started on high-dose corticosteroids and shows marked improvement over the course of the following week.

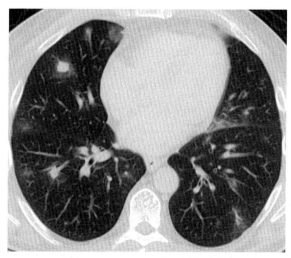

FIGURE 24.1 Chest computed tomography showing bilateral patchy pulmonary infiltrates and ground-glass nodules. *Courtesy of Tan-Lucien Mohammed, MD.*

Rare and Interesting Cases in Pulmonary Medicine.
Copyright © 2017 Elsevier Inc. All rights reserved.

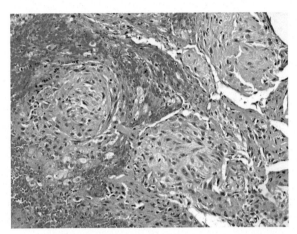

FIGURE 24.2 Lung histology revealing organizing pneumonia surrounded by alveolar fibrin. *Courtesy of Mitra Mehrad, MD.*

What is the diagnosis?

ACUTE FIBRINOUS AND ORGANIZING PNEUMONIA

Acute fibrinous and organizing pneumonia (AFOP) is a distinct histological form of lung injury characterized by the presence of intraalveolar fibrin and organizing pneumonia (Fig. 24.2). It was first described by Beasley in 2002, and since then it has been better described and characterized.

Two distinct clinical patterns of AFOP exist: an acute, rapidly progressive form with a high mortality and a milder subacute form that responds positively to treatment with a good chance of recovery. The exact etiology of AFOP remains unclear; it may be idiopathic or occur in association with various drugs (amiodarone, abacavir, and decitabine), autoimmune diseases, lymphoma, hematopoietic stem call transplantation, environmental exposures, and infections.

Patients commonly present with fevers, nonproductive cough, chest pain, hemoptysis, and progressive dyspnea. Radiological pattern on CT consists of diffuse, bilateral, patchy, pulmonary infiltrates, and ground-glass opacities, in a peripheral basilar distribution. Pulmonary function tests may show a mixed restrictive and obstructive ventilatory defect and bronchoalveolar lavage is often nondiagnostic.

Diagnosis is made on lung biopsy. Histology shows prominent intraalveolar deposits of fibrin (termed "fibrin balls") and organizing pneumonia. There is absence of hyaline membranes and eosinophils, which differentiate it from diffuse alveolar damage and eosinophilic pneumonia, respectively. Varying amounts of type II pneumocyte hyperplasia, inflammation, and edema may also be present.

Since the disease is uncommon, there are no accepted guidelines for its management. The subacute form of the disease has a good prognosis as it usually responds to treatment with corticosteroids or other immunosuppressive medication. The severe acute form of AFOP results in fulminant respiratory failure requiring mechanical ventilation, is poorly responsive to corticosteroids and immunotherapy, and is often fatal.

AFOP is a distinct histologic type of organizing pneumonia. Another is granulomatous organizing pneumonia. It is characterized by radiologic findings of pulmonary nodules or masses and on histology by the presence of nonnecrotizing granulomas and organizing pneumonia.

TAKEAWAY POINTS

- AFOP is a histological diagnosis based on the finding of alveolar fibrin and organizing pneumonia on lung biopsy.
- There are two forms of the disease: an acutely progressive type with a high mortality and a subacute variant that is responsive to immunosuppressive therapy.

FURTHER READING

Beasley, M.B., Franks, T.J., Galvin, J.R., Gochuico, B., Travis, W.D., 2002. Acute fibrinous and organizing pneumonia: a histological pattern of lung injury and possible variant of diffuse alveolar damage. Arch. Pathol. Lab. Med. 126 (9), 1064–1070.

Feinstein, M.B., DeSouza, S.A., Moreira, A.L., Stover, D.E., Heelan, R.T., et al., 2015. A comparison of the pathological, clinical and radiographical, features of cryptogenic organising pneumonia, acute fibrinous and organising pneumonia and granulomatous organising pneumonia. J. Clin. Pathol. 68 (6), 441–447.

Garcia, B.A., Goede, T., Mohammed, T.L., 2015. Acute fibrinous organizing pneumonia: a case report and literature review. Curr. Probl. Diagn. Radiol. 44 (5), 469–471.

Case 25

A 31-year-old female with a history of systemic lupus erythematosus (SLE) presents with a 4-month history of progressive shortness of breath on exertion. She was diagnosed with SLE over 10 years ago and is currently on low-dose prednisone and methotrexate. She has a history of pleural effusions and recurrent pleuritic chest pain episodes in the lower anterior and lateral aspects of her chest. She denies any fevers, chills, cough, or wheezing.

Her vital signs and physical examination are unremarkable except for reduced breath sounds at the lung bases. Her chest X-ray shows low lung volumes but no parenchymal abnormalities. Her pulmonary function tests (PFTs) results compared to a year ago are shown in Table 25.1. High-resolution computed tomography of the chest shows normal lung parenchyma and her echocardiography showed normal cardiac function.

TABLE 25.1 PFTs (Shown as Percentage of Predicted) on Presentation Compared to a Year Ago

Parameter	Current Presentation	One Year Prior
FVC (%)	52	90
FEV1 (%)	48	89
FEV1/FVC	92	99
TLC (%)	67	86
DLCO (%)	60	90
DLCO/VA (%)	88	90

DLCO, diffusing capacity of the lung for carbon monoxide; *FEV1*, forced expiratory volume in 1 s; *FVC*, forced vital capacity; *PFTs*, pulmonary function tests; *TLC*, total lung capacity; *VA*, alveolar volume.

What is the diagnosis?

SHRINKING LUNG SYNDROME

The lungs are commonly affected in SLE in various ways. Shrinking lung syndrome (SLS) is one of the rare manifestations that occur in around 0.5% of cases. SLS was first described by Hoffbrand in 1965 and is characterized by progressive lung volume loss and a restrictive ventilatory defect in the absence of any intrinsic parenchymal process.

The etiology of SLS is uncertain. Since 65% of patients with SLS have pain secondary to pleural inflammation, the resultant inability to take in a deep inspiration may result in diaphragmatic dysfunction via reflex inhibition of diaphragmatic activation. Another theory is that SLE causes primary myositis or myopathy of the diaphragm muscle. SLS may occur at any stage of the SLE course, but is reported to occur more often in severe cases of SLE that are refractory to traditional treatment. SLS may also occur in Sjogren disease and systemic sclerosis.

Patients usually present with chest pain, shortness of breath on exertion, and orthopnea. Physical examination of the chest is usually unremarkable, except for findings of low lung volumes and sometimes paradoxical abdominal movement. On imaging, low lung volumes and elevated hemidiaphragms are noted with the absence of any other pulmonary pathology. PFTs reveal a restrictive ventilatory defect. Maximal inspiratory pressures and maximal expiratory pressures are reduced, and diffusion capacity for carbon monoxide is low, but normalizes when corrected for alveolar volume, indicating an extrinsic etiology for the restrictive ventilatory defect.

The use of corticosteroids, along with the initiation or augmentation of immunosuppressive therapy, has been used to manage patients with SLS. Azathioprine, mycophenolate, cyclophosphamide, and rituximab have been used with varying success. Theophylline and β-agonists use has also been reported, with the idea that it can augment diaphragmatic function. In a few cases, progressive disease may require the use of noninvasive positive pressure ventilation or invasive mechanical ventilation but death as a result of respiratory failure has rarely been described.

Other pulmonary manifestations of SLE include:

- Pleuritis
- Pleural effusions
- Interstitial lung disease
- Lupus pneumonitis
- Diffuse alveolar hemorrhage
- Pulmonary arterial hypertension
- Pulmonary thromboembolism

TAKEAWAY POINTS

- SLS is characterized by progressive lung volume loss and restrictive ventilatory defect in patients with SLE in the absence of any parenchymal pathology.
- Management with corticosteroids and other immunosuppressive therapies has been reported with varying success.

FURTHER READING

Carmier, D., Marchand-Adam, S., Diot, P., Diot, E., 2010. Respiratory involvement in systemic lupus erythematosus. Revue des Maladies Respiratoires 27 (8), e66–e78.

Hoffbrand, B.I., Beck, E.R., 1965. "Unexplained" dyspnoea and shrinking lungs in systemic lupus erythematosus (Clinical Research ed) BMJ 1 (5445), 1273–1277.

Toya, S.P., Tzelepis, G.E., 2009. Association of the shrinking lung syndrome in systemic lupus erythematosus with pleurisy: a systematic review. Sem. Arthritis Rheum. 39 (1), 30–37.

Case 26

A 43-year-old female with a known history of chronic Hepatitis C cirrhosis and membranous glomerulonephritis is admitted with shortness of breath on exertion, a nonproductive cough, low-grade fevers, arthralgia, weakness, and abdominal pain for over a week. She denies any sputum production, hemoptysis, or any recent sick contacts. Examination is relevant for bilateral inspiratory crackles on lung auscultation and the presence of palpable purpura on the abdomen and all extremities.

A chest X-ray shows bilateral diffuse pulmonary infiltrates. Her dyspnea and hypoxia progress and she requires invasive mechanical ventilation. A bronchoscopy with serial lavages (BAL) is significant for progressively bloody return. On cytology of the BAL fluid, there are abundant hemosiderin-laden macrophages, consistent with diffuse alveolar hemorrhage (DAH). No microorganisms were identified.

Work-up includes negative anti-glomerular basement membrane antibodies, antineutrophil antibodies, and antineuronal nuclear antibodies. Complement levels were reduced and cryoglobulin level is elevated at >0.05 mg/mL.

What is the etiology of the patient's DAH?

CRYOGLOBULINEMIA-ASSOCIATED DIFFUSE ALVEOLAR HEMORRHAGE

Cryoglobulinemia is a known complication of chronic Hepatitis C virus infection. In the absence of any other etiology, and the elevated cryoglobulin levels, the patient in this case had cryoglobulinemia-associated DAH.

Cryoglobulins are immune complexes that precipitate from the blood on exposure to cold temperatures ($<37°C$), and redissolve on rewarming. Cryoglobulinemic vasculitis is a small vessel vasculitis where cryoglobulins deposit in the blood vessel endothelium resulting in the activation of the complement pathway and endothelial damage.

There are three types of cryoglobulins:

- Type I cryoglobulins. These are associated with hematological malignancies.
- Type II cryoglobulins. Occur in the setting of chronic Hepatitis C infection.
- Type III mixed cryoglobulins. They carry rheumatoid factor activity and are associated with Hepatitis C, chronic inflammatory diseases, autoimmune disorders, and lymphoproliferative conditions.

Most patients with type I cryoglobulinemia are asymptomatic. Patients with mixed cryoglobulinemia usually present with palpable purpura, generalized weakness, renal disease, and arthralgia or arthritis. In a small number of cases, severe disease with multi-organ involvement may occur. Systemic manifestations include glomerulonephritis, low complement levels, pulmonary symptoms, and nervous system involvement, usually peripheral neuropathy. Symptomatic pulmonary manifestations are rare and occur in the setting of type II or type III cryoglobulinemia. These may include interstitial lung disease in 2%, DAH in 3.2%, organizing pneumonia, and/or subacute alveolitis. DAH usually occurs in the setting of coexistent membranoproliferative glomerulonephritis.

There is no standard treatment for the pulmonary disease associated with cryoglobulinemia. The presence of DAH carries a poor prognosis and high mortality (80%). Case reports have described the use of various therapies including high-dose corticosteroids, immunosuppressive agents, antiviral agents, and plasmapheresis, usually with little success. Patients frequently succumb to repeated episodes of DAH.

TAKEAWAY POINTS

- Cryoglobulinemia is categorized into three types, with Hepatitis C virus infection being the main cause of type II and type III cryoglobulinemia.
- Pulmonary manifestations including DAH and interstitial lung disease, while uncommon, carry a poor prognosis.

FURTHER READING

Amital, H., Rubinow, A., Naparstek, Y., 2005. Alveolar hemorrhage in cryoglobulinemia–an indicator of poor prognosis. Clin. Exp. Rheumatol. 23 (5), 616–620.

Bombardieri, S., Paoletti, P., Ferri, C., Di Munno, O., Fornal, E., et al., 1979. Lung involvement in essential mixed cryoglobulinemia. Am. J. Med. 66 (5), 748–756.

Ferri, C., Zignego, A.L., Pileri, S.A., 2002. Cryoglobulins. J. Clin. Pathol. 55 (1), 4–13.

Retamozo, S., Diaz-Lagares, C., Bosch, X., Bove, A., Brito-Zeron, P., et al., 2013. Life-threatening cryoglobulinemic patients with hepatitis C: clinical description and outcome of 279 patients. Medicine 92 (5), 273–284.

Case 27

A 52-year-old female is evaluated after being admitted to the hospital complaining of progressive shortness of breath over the last 4 months. Two recent syncopal episodes prompted her admission. She has no known cardiac or pulmonary disease. An echocardiogram shows normal left ventricle function and ejection fraction but elevated right ventricle systolic pressures and an enlarged and hypertrophied right ventricle.

The patient undergoes a right heart catheterization (RHC) with findings of mean pulmonary arterial pressure (mPAP) of 42 mm Hg, pulmonary artery occlusion pressure (PAOP) 10 mm Hg, cardiac output of 2.3 L/min, and pulmonary vascular resistance (PVR) of 14 Wood units. A lung ventilation-perfusion scan is normal, and chest computed tomography (CT) scan shows normal lung parenchyma with some mediastinal adenopathy. Pulmonary function testing reveals normal spirometry and lung volumes with an isolated reduction in diffusing capacity (36% of predicted).

The patient is diagnosed with idiopathic pulmonary arterial hypertension (PAH) after a thorough workup. She is admitted to the intensive care unit and started on intravenous prostacyclin. Shortly after starting the therapy, she develops acute shortness of breath and her chest X-ray reveals pulmonary edema. A CT scan of the chest is shown in Fig. 27.1.

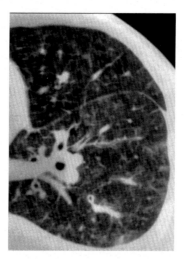

FIGURE 27.1   Axial chest computed tomography in lung windows showing smooth interlobular and airway thickening. *Courtesy of Tan-Lucien Mohammed, MD.*

What is the diagnosis?

Rare and Interesting Cases in Pulmonary Medicine.
Copyright © 2017 Elsevier Inc. All rights reserved.

PULMONARY VENO-OCCLUSIVE DISEASE

Pulmonary veno-occlusive disease (PVOD) is a rare form of PAH that shares many features with other types of PAH but retains a unique histopathological pattern. As such, it is classified under a distinct subgroup category termed Group 1' PAH, alongside pulmonary capillary hemangiomatosis (PCH).

The etiology of PVOD remains to be determined, but several risk factors are associated with the development of the disease, including genetic predisposition (EIF2AK4 gene), exposure to chemotherapeutic drugs and alkylating agents (mitomycin, cyclophosphamide, bleomycin, bis-chloroethylnitrosourea), autoimmune diseases (scleroderma, sarcoidosis), radiation, infections, inhalation of organic solvents, and hematopoietic stem cell transplant.

Presenting symptoms and physical examination findings in PVOD are similar to other forms of PAH. Symptoms are nonspecific and include fatigue, shortness of breath, syncope, and chest pain. Findings on physical examination may include a loud P2, crackles on lung auscultation, and lower extremity edema. PVOD is more rapidly progressive than idiopathic PAH and carries a worse prognosis, with a high mortality within 2 years of diagnosis.

Echocardiography will reveal elevated right ventricle systolic pressures and evidence of right heart strain. RHC will show hemodynamic profile of precapillary pulmonary hypertension with mPAP ≥25 mm Hg, PAOP ≤15 mm Hg, and a PVR ≥3 Wood units. Though the pulmonary capillary pressures are elevated in PVOD, the PAOP is low since it reflects the pressures in the distal large veins and not the pulmonary capillaries.

CT chest scan findings of centrilobular ground-glass opacities, prominent septal thickening, mosaic attenuation, mediastinal lymph node enlargement, and small pleural effusions can distinguish PVOD from PAH. Patients may also have hemosiderin-laden macrophages on bronchoalveolar lavage (BAL), a low diffusing capacity of the lungs for carbon monoxide ≤55%, and an arterial oxygenation ≤70 mm Hg. A ventilation-perfusion scan is often normal or may show patchy perfusion defects.

Though definitive diagnosis would require histological examination of the lung, open lung biopsy carries significant risk in these patients, and noninvasive diagnosis can usually be made based upon clinical exam, imaging, hemodynamics, and response to therapy. Testing for the EIF2AK4 mutation in patients with presumed PVOD can also help establish the diagnosis. Histology of lung tissue, obtained at lung transplantation or autopsy, shows fibrous intimal proliferation of the pulmonary venules and small veins with central recanalization of the obstructed vessels (Fig. 27.2). Passive congestion of the upstream capillaries results in a dilated "loop-like" appearance proximal to the occluded vessels and occult hemorrhage; thus the finding of hemosiderin-laden macrophages in BAL fluid obtained from patients with PVOD.

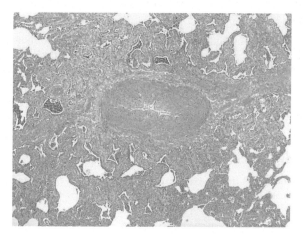

FIGURE 27.2   Evidence of marked myointimal thickening and sclerosis of the pulmonary vein within the interlobular septum. There is also evidence of alveolar hemosiderosis indicating prior bleeding into the alveolar spaces. *Courtesy of Humberto Trejo Bittar, MD.*

Patients with PVOD have a poor response to PAH-specific therapy and may develop noncardiogenic pulmonary edema upon starting treatment. Patients should be promptly referred to experienced centers for management. Lung transplantation at this time is the only curative therapy.

PULMONARY CAPILLARY HEMANGIOMATOSIS

PCH is characterized by the proliferation of pulmonary capillaries in the alveolar septal walls resulting in pulmonary hypertension. These proliferating capillaries infiltrate the interstitium of the lung, the walls of small pulmonary vessels, bronchi, and pleura. PCH falls in the same category of Group 1' PAH as PVOD. The clinical presentation, examination findings, and workup are identical to PVOD. The etiology of PCH remains unknown, although it may be seen in association with connective tissue diseases. A hereditary form, with mutations in the autosomal recessive gene EIF2AK4, has been described.

Although regarded as different disease entities, PCH and PVOD may in fact represent a continuum of the same disease process. The dilated capillaries secondary to the passive congestion in the setting of distal venous obstruction in PVOD may give rise to PCH. Over 74% of patients with PVOD have reported concomitant histological findings of PCH, while all patients with PCH have histological findings of PVOD. As in PVOD, patients with PCH responded poorly to PAH-specific therapy and are at risk of developing acute pulmonary edema with vasodilator therapy. Given the poor prognosis of PCH, most patients with this disease should be considered for lung transplantation.

TAKEAWAY POINTS

- PVOD is a type of Group 1' PAH involving the small pulmonary veins and venules. It is often misdiagnosed as idiopathic PAH due to similar presentation and RHC hemodynamic findings.
- PAH patients with characteristic chest CT findings, and who acutely deteriorating upon staring PAH-specific therapy should be suspected of having PVOD.

FURTHER READING

Eyries, M., Montani, D., Girerd, B., Perret, C., Leroy, A., Lonjou, C., et al., 2014. EIF2AK4 mutations cause pulmonary veno-occlusive disease, a recessive form of pulmonary hypertension. Nat. Genet. 46 (1), 65–69.

Huertas, A., Girerd, B., Dorfmuller, P., O'Callaghan, D., Humbert, M., Montani, D., 2011. Pulmonary veno-occlusive disease: advances in clinical management and treatments. Expert Rev. Respir. Med. 5 (2), 217–229.

Lantuejoul, S., Sheppard, M.N., Corrin, B., Burke, M.M., Nicholson, A.G., 2006. Pulmonary veno-occlusive disease and pulmonary capillary hemangiomatosis: a clinicopathologic study of 35 cases. Am. J. Surg. Pathol. 30 (7), 850–857.

Simonneau, G., Gatzoulis, M.A., Adatia, I., Celermajer, D., Denton, C., Ghofrani, A., et al., 2013. Updated clinical classification of pulmonary hypertension. J. Am. Coll. Cardiol. 62 (25 Suppl.).

Case 28

A 40-year-old nonsmoking female presents with chronic nonproductive cough for the last 3 years. She has a history of surgically treated breast cancer 10 years prior. She denies any other symptoms and her physical examination is unremarkable. High-resolution computed tomography (HRCT) of chest shows small bilateral nodules and areas of mosaic attenuation. Bronchoscopy with transbronchial biopsies is unrevealing, so the patient undergoes video-assisted thoracoscopic surgery lung biopsy. Histological analysis shows neuroendocrine cell hyperplasia (Fig. 28.1) with positive staining for CD56 and chromogranin.

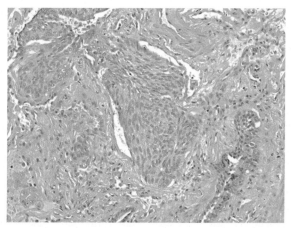

FIGURE 28.1 A tumorlet composed of oval to spindle-shaped cells with moderate amount of eosinophilic cytoplasm and salt and pepper chromatin. *Courtesy of Mitra Mehrad, MD.*

What is the diagnosis?

DIFFUSE IDIOPATHIC PULMONARY NEUROENDOCRINE CELL HYPERPLASIA

Diffuse idiopathic pulmonary neuroendocrine cell hyperplasia (DIPNECH) was first described in the 1950s, but the name was not coined until 1992. This disorder of idiopathic pulmonary neuroendocrine cell (PNEC) hyperplasia occurs most frequently in middle-aged, nonsmoking females and may be associated with peripheral carcinoid tumors. DIPNECH may take many forms in the lungs and is classified by the World Health Organization as either:

1. Generalized proliferation of PNEC
2. Small nodules (neuroendocrine bodies)
3. Liner proliferation of PNEC, confined to the respiratory epithelium

   PNEC is confined to the bronchial and bronchiolar epithelium. PNEC hyperplasia often is reactive secondary to airway inflammation or fibrosis; however, in DIPNECH, the PNEC hyperplasia is not associated with any inciting factors. Once the neuroendocrine cells invade beyond the basement membrane, they are termed tumorlets if their diameter is ≤5 mm, or carcinoid tumors if the nodule diameter is >5 mm.

   Patients may present with nonproductive cough, wheezing, and dyspnea or be referred for evaluation of incidentally found pulmonary nodule or nodules. Pulmonary function tests may reveal an obstructive ventilatory defect or a mixed obstructive/restrictive profile. On HRCT of the chest, the most common pattern is that of multiple, bilateral, reticulonodular infiltrates, ground-glass opacities, bronchiectasis, and mosaic attenuation with air trapping.

   Diagnosis is established on histopathological evaluation of surgical lung biopsy. There will be evidence of PNEC hyperplasia confined to the airway epithelium along with tumorlet formation. Tumor immunohistochemistry staining for neuroendocrine markers (chromogranin, synaptophysin, CD56, CK7, and neuron-specific enolase) is positive. Transbronchial biopsy samples are often inadequate to establish the diagnosis.

   Histological association with carcinoid tumors and changes of constrictive bronchiolitis are often found. Serum chromogranin A as a tumor marker was found to be elevated in approximately half of patients in one study with no correlation with the number or size of tumors. Urine 5-hydroxyindoleacetic acid analysis is often normal.

   There is no agreement on optimal therapy. Various management strategies have been reported including observation in mild cases, surgical resection, inhaled and systemic corticosteroids, chemotherapy, bronchodilators, and somatostatin analogs. There is debate on whether somatostatin analogs improve symptoms or alter the course of the disease. The majority of patients have a stable clinical course, with a few progressing to respiratory failure. Patients with severe progressive disease may be referred for lung transplantation.

TAKEAWAY POINTS

- DIPNECH should be suspected in middle-aged, nonsmoking females with nonspecific pulmonary symptoms and reticulonodular pulmonary infiltrates.
- The disease follows a slow indolent course with a generally favorable prognosis.

FURTHER READING

Aguayo, S.M., Miller, Y.E., Waldron Jr., J.A., Bogin, R.M., Sunday, M.E., Staton Jr., G.W., et al., 1992. Brief report: idiopathic diffuse hyperplasia of pulmonary neuroendocrine cells and airways disease. N. Engl. J. Med. 327 (18), 1285–1288.

Beasley, M.B., Brambilla, E., Travis, W.D., 2005. The 2004 World Health Organization classification of lung tumors. Semin. Roentgenol. 40 (2), 90–97.

Chauhan, A., Ramirez, R.A., 2015. Diffuse idiopathic pulmonary neuroendocrine cell hyperplasia (DIPNECH) and the role of somatostatin analogs: a case series. Lung 193 (5), 653–657.

Felton 2nd, W.L., Liebow, A.A., Lindskog, G.E., 1953. Peripheral and multiple bronchial adenomas. Cancer 6 (3), 555–567.

Nassar, A.A., Jaroszewski, D.E., Helmers, R.A., Colby, T.V., Patel, B.M., Mookadam, F., 2011. Diffuse idiopathic pulmonary neuroendocrine cell hyperplasia: a systematic overview. Am. J. Respir. Crit. Care Med. 184 (1), 8–16.

Wirtschafter, E., Walts, A.E., Liu, S.T., Marchevsky, A.M., 2015. Diffuse idiopathic pulmonary neuroendocrine cell hyperplasia of the lung (DIPNECH): current best evidence. Lung 193 (5), 659–667.

Case 29

A 52-year-old male presents with exertional shortness of breath for the last 4 months as well as bilateral lower extremity pain without swelling. He denies cough, wheezing, hemoptysis, or chest pain. He has no history of smoking or drug abuse, and no personal or family history of cancer or lung disease. High-resolution computed tomography (HRCT) of his chest shows evidence of reticulonodular thickening, cystic changes, and periaortic soft tissue thickening (Fig. 29.1). Imaging of his lower extremities reveals bilateral symmetrical sclerotic lesions of the distal femur and proximal tibia. Biopsy of the sclerotic bone region revealed non-Langerhans cells that stained CD68 positive and CD1a negative.

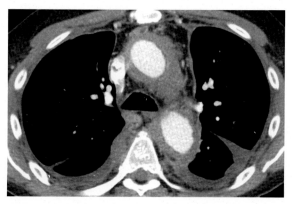

FIGURE 29.1 Chest computed tomography scan with contrast demonstrating periaortic soft tissue thickening giving a "coated aorta" appearance. *Courtesy of Tan-Lucien Mohammed, MD.*

What is the diagnosis?

Rare and Interesting Cases in Pulmonary Medicine.
Copyright © 2017 Elsevier Inc. All rights reserved.

ERDHEIM-CHESTER DISEASE

First described by William Chester in 1930, Erdheim-Chester disease (ECD) is a rare systemic disease characterized by xanthogranulomatous infiltration of the bones and organs by non-Langerhans foamy histiocytes. It commonly presents in adults in their 50s to 70s with a slight male predominance.

The etiology of ECD remains unclear, but dysregulation of chemokines in a predisposed individual (with resultant tissue infiltration by histiocytes) may play a major role. Histological staining in ECD is positive for CD68 but negative for Langerhans cells, CD1a, and S100. Cases of overlap between ECD and Langerhans cell disease have been described in the literature. Most patients with ECD are asymptomatic and bone pain is the most frequent presenting symptom.

Pulmonary involvement is uncommon but is a major cause of morbidity and mortality in these patients. Symptomatic patients will often complain of a non-productive cough and shortness of breath. Pulmonary function tests often show a restrictive ventilatory defect. HRCT of the chest may have findings of reticulonodular infiltrates, cystic lesions, micronodules, interstitial fibrosis, ground-glass opacities, pleural thickening, and pleural effusions.

The systemic manifestations include:

- Constitutional symptoms: fever, night sweats, or weight loss
- Skin: xanthomas
- Skeletal (90%): bilateral symmetrical cortical sclerosis of the diaphysis and metaphysis of long bones, particularly in the lower limbs. Technetium-99m bone scan shows increased uptake in affected regions.
- Central nervous system (CNS; 55%): diabetes insipidus, cerebellar ataxia, visual disturbances, or exophthalmos
- Renal: perirenal fat infiltration giving rise to "hairy kidneys" appearance on computed tomography (CT)
- Cardiac: pseudotumor lesion of the right atrium, pericardial effusion, cardiac infiltration, circumferential infiltration of the aorta giving rise to "coated aorta" appearance on contrasted CT (present in 80% of those with cardiac involvement)
- Retroperitoneal fibrosis: usually spares the inferior vena cava and pelvic ureters, a feature seen in idiopathic retroperitoneal fibrosis

Diagnosis is established by findings on histology of foamy non-Langerhans histiocytes (Touton-like giant cells) infiltrating the tissue, with immunostaining positive for CD68 and negative for CD1a. Pulmonary, cardiac, and CNS involvement portend a worse prognosis. No standard guidelines on managing ECD yet exist. The efficacy of steroids and other immunomodulator drugs have been limited. However, recent reports of therapy with an interferon-a, IL-1R antagonist (Anakinra), and inflixmab have been encouraging; BRAF V600E mutations have been identified in half of patients with ECD, allowing for new treatment with the BRAF inhibitor vemurafenib, with high success rates.

TAKEAWAY POINTS

- ECD is a non-Langerhans histiocytosis with pulmonary manifestations similar to LCH.
- Imaging findings that suggest the diagnosis include symmetrical osteosclerosis of the long bones, hairy kidneys, aortic coating, and right atrial pseudotumor.
- ECD immunostaining is positive for CD68 but negative for CD1a and S100.

FURTHER READING

Chester, W., 1930. Über lipoidgranulomatose. Virchows Arch. Pathol. Anat. 279, 561–602.

Haroche, J., Cohen-Aubart, F., Emile, J.F., Maksud, P., Drier, A., Toledano, D., et al., 2015. Reproducible and sustained efficacy of targeted therapy with vemurafenib in patients with BRAF(V600E)-mutated Erdheim-Chester disease. J. Clin. Oncol. 33 (5), 411–418.

Mazor, R.D., Manevich-Mazor, M., Shoenfeld, Y., 2013. Erdheim-Chester disease: a comprehensive review of the literature. Orphanet J. Rare Dis. 8, 137.

Veyssier-Belot, C., Cacoub, P., Caparros-Lefebvre, D., Wechsler, J., Brun, B., Remy, M., et al., 1996. Erdheim-Chester disease. Clinical and radiologic characteristics of 59 cases. Medicine 75 (3), 157–169.

Case 30

A 52-year-old male was evaluated for shortness of breath 3 days after undergoing extracorporeal shock wave lithotripsy for renal stones. His vital signs were stable, and on examination he had dullness to percussion and decreased breath sounds over the right lung. A chest X-ray showed a large right-sided pleural effusion. The patient underwent a thoracentesis, and had a 1.5 L of straw-colored fluid removed with significant improvement in his dyspnea. Pleural fluid analysis revealed a glucose of 58 mg/dL, pH 7.0, lactate dehydrogenase (LDH) 132 U/L, protein 2.8 g/dL, and creatinine of 4.3 mg/dL. The patient's serum LDH was 213 U/L, protein 6.4 g/dL, and creatinine 1.3 mg/dL.

What is the diagnosis?

URINOTHORAX

Urinothorax is a rare and often underrecognized cause of pleural effusions, characterized by the presence of urine in the pleural space. It was first described by Corriere in 1969 and since has been reported to occur secondary to malignancy, obstructive uropathy, trauma, and after a urological or obstetric procedure.

The pleural effusion occurs as a result of disruption of the urinary tract leading to leakage of urine into the retroperitoneal space and from there into the pleural space through anatomical defects in the diaphragm. Effusions are often unilateral on the same side as the urinary flow disruption or obstruction, but contralateral effusions have been reported. Unilateral obstructions do not cause urinothorax as along as the contralateral kidney remains functioning. Thus bilateral obstruction at the level of the bladder or urethra or physical disruption of the urinary system is required to form a urinothorax. As such, urinothorax has been classified into two categories:

1. Obstructive urinothorax, as a result of bilateral obstructive uropathy
2. Traumatic/iatrogenic urinothorax

Patients may be asymptomatic with the effusion being detected incidentally on imaging. Symptomatic patients may complain of shortness of breath, chest discomfort, and fevers.

The effusions accumulate rapidly and on analysis are transudates by Light's criteria with a pleural pH often below 7.30 and a low pleural glucose. In some cases, the effusion may have a distinctive urine smell. Diagnosis is established based on finding of a pleural creatinine to serum creatinine ratio >1.0. The use of renal scintigraphy with a radioactive tracer can also help confirm the diagnosis and establish the location of defect.

Management consists of correcting the underlying cause, for example relieving the urinary obstruction or surgically repairing tears in the urinary system. Thoracentesis may be sufficient for diagnosis and relief of symptoms in patients with small effusions, but chest tube placement may be required in patients with severe dyspnea or hypoxia. Pleurodesis is rarely indicated.

TAKEAWAY POINTS

- Urinothorax should be suspected in the setting of pleural effusion after a renal procedure.
- Urinothorax is commonly transudate by Light's criteria with a pleural to serum creatinine ratio >1.0.

FURTHER READING

Corriere Jr., J.N., Miller, W.T., Murphy, J.J., 1968. Hydronephrosis as a cause of pleural effusion. Radiology 90 (1), 79–84.

Garcia-Pachon, E., Romero, S., 2006. Urinothorax: a new approach. Curr. Opin. Pulm. Med. 12 (4), 259–263.

Stark, D.D., Shanes, J.G., Baron, R.L., Koch, D.D., 1982. Biochemical features of urinothorax. Arch. Intern. Med. 142 (8), 1509–1511.

Case 31

A 31-year-old male presents with intermittent fevers, a nonproductive cough, and shortness of breath on exertion that started 3 weeks ago. He is a nonsmoker, has no known medical problems, and does not partake in recreational drugs. He works as an accountant and does not recall any recent significant exposures to birds, feathered duvets, hay, or hot tubs. He does report that there was a flood in the basement of his home 1 month ago, and he has been working in the basement, cleaning and fixing the plumbing for the last few weeks. Pulmonary function tests show mixed obstructive and restrictive ventilatory defects with mild reduction in diffusing capacity of the lungs for carbon monoxide (DLCO). A chest computed tomography scan is shown (Fig. 31.1). The patient undergoes a bronchoscopy with bronchoalveolar lavage (BAL) and the BAL fluid is lymphocyte predominant (48%) with a low CD4+ to CD8+ ratio.

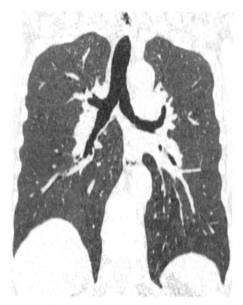

FIGURE 31.1   Chest computed tomography showing bilateral mosaic attenuation with centrilobular ground-glass nodules throughout the lungs. *Courtesy of Tan-Lucien Mohammed, MD.*

What is the diagnosis?

HYPERSENSITIVITY PNEUMONITIS

Hypersensitivity pneumonitis (HP), or extrinsic allergic alveolitis, is a form of granulomatous interstitial lung disease characterized by repeated inflammatory episodes in response to exposure to specific inhaled organic or inorganic particles. The first cases were described in farmers exposed to moldy hay and straw, giving rise to the term "farmer's lung." The disease manifests as an acute, a subacute, or a chronic form depending on the intensity and duration of the exposure. In our case, the patient had basement lung HP, likely related to exposure to *Cephalosporium* or *Penicillium* species in the contaminated flooded basement.

Patients often present with complaints of dyspnea and a nonproductive cough, but may also have fevers, chills, weight loss, chest pain, and arthralgia. On examination, inspiratory crackles may be heard on lung auscultation. Diagnosis is often overlooked, as symptoms are nonspecific and a history of relevant exposures is not obtained. As such, the importance of obtaining a detailed and thorough exposure, occupational, and social history cannot be stressed enough as this may determine the inciting factor. Acute HP may manifest within hours of exposure with viral-like symptoms, whereas subacute HP occurs over days to weeks. The chronic form may develop and manifest over several months. It has been observed that the disease is more common in nonsmokers, possibly related to the inhibitory effects of nicotine.

Myriad causes have been implicated in the etiology of HP, which include exposure to avian feathers (parakeets), hot tubs, molds, and chemical compounds (Table 31.1). However, despite a detailed history and workup, no causative agent is found in approximately 30% of cases. The general population is continuously exposed to HP-related antigens yet do not develop HP, although they might have positive precipitin antibody tests to antigens. This raises the theory that HP develops on exposure to environmental antigens in genetically predisposed individuals.

High-resolution computed tomography of the lungs in acute and subacute HP may show areas of mosaic attenuation and centrilobular ground-glass nodules with upper and middle lung zones predominance. In chronic or "burned-out" HP, a fibrotic interstitial pattern predominates similar to usual interstitial pneumonia or fibrotic nonspecific interstitial pneumonia. Pulmonary function tests often show a restrictive ventilatory defect with a reduced DLCO. A bronchoscopy with BAL is frequently done as part of the evaluation, and typically reveals a lymphocytic alveolitis (>30% lymphocytes) with a CD4+ to CD8+ ratio of less than 1. In contrast, sarcoidosis is associated with a high CD4+ to CD8+ ratio. The serum precipitating immunoglobulin-G antibodies against offending antigens can be measured, and are often detected in patients with HP. However, the same antibodies may also be found in exposed yet asymptomatic individuals, making their utility in the diagnosis of HP low. On histology, there is the presence of poorly defined granulomas, bronchiolitis, and alveolitis (Fig. 31.2).

TABLE 31.1 Some of the Various Types of HP

Bird fancier's lung; pigeon breeder's lung

Feather duvet lung

Farmer's lung

Hot tub lung

Humidifier lung

Summer type pneumonitis

Machine operator's lung

Chemical worker's lung

Bagassosis (moldy sugarcane)

Animal handler's and laboratory worker's lung

Familial HP

Suberosis (moldy cork)

Basement lung

Wine maker's lung

Wind instrument lung

HP, hypersensitivity pneumonitis.

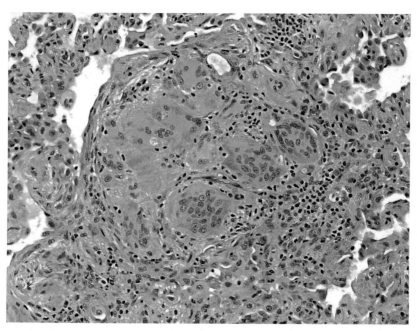

FIGURE 31.2   Lung parenchyma with bronchiolocentric nonnecrotizing poorly formed granulomas with lymphocytic inflammation. *Courtesy of Humberto Trejo Bittar, MD.*

Avoidance of exposure to causative agent is the cornerstone of treatment along with treatment with corticosteroids. In cases where the etiology cannot be determined and there is lack of response to corticosteroids, treatment with other immunosuppressive therapies can be attempted. Patients who have advanced fibrosis due to prolonged exposure and inflammation as a result of chronic HP may require referral for lung transplantation.

TAKEAWAY POINTS

- HP is a granulomatous interstitial lung disease that occurs in response to inhaled organic and inorganic antigens in predisposed individuals.
- A high degree of suspicion and thorough environmental and occupational history are required to determine inciting factor. No etiology is found in up to 30% of cases.
- Avoidance of precipitating agent and corticosteroids are mainstay of treatment in acute and subacute form.

FURTHER READING

Blanchet, M.R., Israel-Assayag, E., Cormier, Y., 2004. Inhibitory effect of nicotine on experimental hypersensitivity pneumonitis in vivo and in vitro. Am. J. Respir. Crit. Care Med. 169 (8), 903–909.

Campbell, J.M., 1932. Acute symptoms following work with hay. Br. Med. J. 2, 1143–1144.

Costabel, U., Bonella, F., Guzman, J., 2012. Chronic hypersensitivity pneumonitis. Clin. Chest Med. 33 (1), 151–163.

Lacasse, Y., Selman, M., Costabel, U., Dalphin, J.C., Ando, M., Morell, F., et al., 2003. Clinical diagnosis of hypersensitivity pneumonitis. Am. J. Respir. Crit. Care Med. 168 (8), 952–958.

Case 32

A 34-year-old female with a history of albinism and gastrointestinal complaints presents with complaints of progressive shortness of breath on exertion for the last year. She is a nonsmoker, has no significant exposures, no significant family medical history, and her parents are originally from Puerto Rico. On examination, she has hypopigmentation of the skin and hair as well as bruising on the skin. Bilateral basal inspiratory crackles are present on lung examination.

Pulmonary function tests (PFTs) are consistent with an intrinsic restrictive ventilatory defect. High-resolution computed tomography (HRCT) of her chest is shown in Fig. 32.1. The patient undergoes a lung biopsy, and her histology reveals fibrotic parenchymal changes with evidence of clear vacuolated type II pneumocytes.

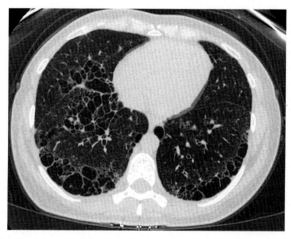

FIGURE 32.1   Chest computed tomography scan showing bilateral lower lobe fibrosis and honeycombing.

What is the diagnosis?

HERMANSKY-PUDLAK SYNDROME

Hermansky-Pudlak syndrome (HPS) is an autosomal recessive disorder that was first described in 1959, and is commonly seen among individuals from northwest Puerto Rico. The syndrome is characterized by the triad of oculocutaneous albinism, platelet dysfunction, and lysosomal accumulation of ceroid lipofuscin. These manifestations occur as a result of abnormal formation and trafficking of intracellular vesicles such as melanosomes, platelet dense granules, and lysosomes, respectively. The ceroid lipofuscin accumulation in lysosomes is believed to be the cause of the systemic manifestations, such as pulmonary fibrosis and granulomatous colitis. Another proposed mechanism for the development of the pulmonary fibrosis in HPS is an inflammatory response to recurrent pulmonary hemorrhage.

There are nine different variants of HPS that have been described to date, distinguishable by their clinical presentation and genetic etiology. Pulmonary fibrosis is associated with only types 1 and 4 of the disease.

Patients with lung involvement may present in their early 30s with progressive shortness of breath and coughing. On examination, there is evidence of albinism as evident by the hypopigmented skin, hair, and iris on transillumination. On lung auscultation, inspiratory crackles may be present.

PFTs will reveal a restrictive ventilatory defect. Chest HRCT will show bilateral, peripherally distributed septal thickening, reticulations, and ground-glass opacities, more predominantly in the middle and lower lung zones. As the disease progress, more fibrotic changes and central involvement occur. On lung biopsy, there is presence of clear vacuolated or foamy type II pneumocytes on a background of parenchymal fibrosis. Examination of the platelets under electron microscopy would reveal absence of dense granules.

The bleeding diathesis is prevented and treated with desmopressin and platelet transfusions. Pulmonary fibrosis accounts for the mortality of 50% of patients by their fifth decade. There are published reports of the use of pirfenidone to slow disease progression but further research is required. Lung transplantation is reserved for advanced cases; however, there have been only a few successful lung transplants reported for HPS due to complications related to the bleeding diathesis.

TAKEAWAY POINTS

- HPS is characterized by oculocutaneous albinism, dysfunctional platelets, and ceroid lipofuscin cell accumulation.
- HPS should be considered in individuals presenting with pulmonary fibrosis and gastrointestinal complaints.

FURTHER READING

Avila, N.A., Brantly, M., Premkumar, A., Huizing, M., Dwyer, A., Gahl, W.A., 2002. Hermansky-Pudlak syndrome: radiography and CT of the chest compared with pulmonary function tests and genetic studies. Am. J. Roentgenol. 179 (4), 887–892.

Brantly, M., Avila, N.A., Shotelersuk, V., Lucero, C., Huizing, M., Gahl, W.A., 2000. Pulmonary function and high-resolution CT findings in patients with an inherited form of pulmonary fibrosis, Hermansky-Pudlak syndrome, due to mutations in HPS-1. Chest 117 (1), 129–136.

Garay, S.M., Gardella, J.E., Fazzini, E.P., Goldring, R.M., 1979. Hermansky-Pudlak syndrome. Pulmonary manifestations of a ceroid storage disorder. Am. J. Med. 66 (5), 737–747.

Hermansky, F., Pudlak, P., 1959. Albinism associated with hemorrhagic diathesis and unusual pigmented reticular cells in the bone marrow: report of two cases with histochemical studies. Blood 14 (2), 162–169.

Case 33

A 28-year-old male with a history of cerebral vein thrombosis presents with fevers, chills, and coughing up blood for the last 3 days. He is a nonsmoker, does not use recreational drugs, and has no significant environmental exposures. His examination is unremarkable with no evidence of skin rashes, mouth or genital ulcers. During his workup for the hemoptysis, he undergoes a contrasted chest computed tomography (CT) scan that shows multiple bilateral pulmonary aneurysms. There are no parenchymal infiltrates and no evidence of pulmonary embolism. Ultrasound examination of the lower extremities reveals bilateral deep vein thrombosis.

Given the constellation of findings, what is the diagnosis?

HUGHES-STOVIN SYNDROME

First described by British physicians Hughes and Stovin in 1959, Hughes-Stovin syndrome is a very rare vasculitic disorder of unclear etiology, believed to be a vascular variant of Behcet disease. It is characterized by the presence of pulmonary artery aneurysms and systemic thrombosis.

The disease usually affects young males in their 20s to 30s, with patients often presenting with fevers, chills, cough, chest pain, and hemoptysis. Mouth and genital ulcers, as well as eye involvement seen in Behcet disease, are absent in patients with Hughes-Stovin syndrome. Neurological symptoms including headache, diplopia, and seizures may occur in the setting of cerebral venous sinus thrombosis. Chest CT reveals findings of single or multiple pulmonary or bronchial artery aneurysms and there may also be evidence of pulmonary artery thrombosis.

The syndrome occurs in three stages:

- Stage 1: thrombophlebitis (patients often will present with fevers and chills)
- Stage 2: the development of pulmonary artery aneurysm
- Stage 3: pulmonary artery aneurysm rupture

Pulmonary artery aneurysm rupture is a dreaded and often fatal complication of this disease, and patients should be monitored closely for pulmonary aneurysm enlargement and new onset hemoptysis.

Management includes treatment with corticosteroids and immunotherapy, particularly cyclophosphamide. With treatment, patients may have regression of their pulmonary artery aneurysms. Those with acute onset of hemoptysis may require embolization of the bleeding vessel. When bleeding is very severe, surgical management with lobectomy or even pneumonectomy may be required. Patients may be treated with anticoagulation therapy for deep vein thrombosis, but should be monitored closely for any evidence of bleeding.

TAKEAWAY POINTS

- Hughes-Stovin syndrome is a vasculitic disorder characterized by pulmonary artery aneurysms and deep vein thrombosis.
- Patients should be managed with corticosteroids and cytotoxic therapy. In some cases, they may require surgical intervention or embolization of pulmonary aneurysms.
- The role of anticoagulation remains controversial.

FURTHER READING

Ataya, A., Alnuaimat, H., 2014. Pulmonary artery aneurysms. Am. J. Respir. Crit. Care Med. 190 (7), e26–e27.

Hughes, J.P., Stovin, P.G., 1959. Segmental pulmonary artery aneurysms with peripheral venous thrombosis. Br. J. Dis. Chest. 53, 19–27.

Khalid, U., Saleem, T., 2011. Hughes-Stovin syndrome. Orphanet J. Rare Dis. 6, 15.

Case 34

A 64-year-old male is evaluated for acute onset of hypoxia that started within 30 min after an elective right total knee arthroplasty. The patient had an uneventful intraoperative course, but shortly after surgery was noted to be tachycardic, tachypneic, hypotensive, and hypoxic. A computed tomography scan of the chest with contrast was negative for pulmonary embolism, but showed patchy bilateral ground-glass opacities. An echocardiogram demonstrated a preserved left ventricle ejection fraction of 55%, elevated right ventricle systolic pressures, and reduced right ventricular function. The patient did not suffer from any neurological derangement and did not have any skin rashes on examination. He is transferred to the intensive care unit and recovers after fluid resuscitation and supportive care.

What is the diagnosis?

BONE CEMENT IMPLANTATION SYNDROME

The acute onset of hypoxia, hypotension, altered mental status, cardiac arrhythmias, and hypotension as a result of femoral canal instrumentation and cement implantation during orthopedic surgery is known as bone cement implantation syndrome (BCIS). It was first described in 1970, and it occurs during or immediately after surgery. It is commonly associated with vertebroplasty, and hip or knee arthroplasty. Risk is highest in elderly patients with cardiovascular disease undergoing cemented prosthetic replacement for fracture repair.

Multiple mechanisms contribute to the hemodynamic instability in BCIS. Prosthetic cement contains methyl methacrylate (MMA), and the release of monomers secondary to high intramedullary pressures forces air and medullary contents into the circulation. This results in the release of histamines, complement activation, and endogenous cannabinoid (anandamide and 2-arachidonoylglycerol)-induced vasodilatation. In reported cases of intraoperative death, autopsy has confirmed the presence of fat, bone marrow, bone emboli, and MMA particles in the lung.

The first indication of BCIS during surgery may be a fall in the end-tidal CO_2 concentration. Reported potential complications of BCIS include acute respiratory distress syndrome, bronchoconstriction, pulmonary edema, hypothermia, and thrombocytopenia. The rate of cardiac arrest and death from BCIS has been reported to be between 0.6% and 4.3%.

BCIS can be graded according to severity:

- Grade 1: moderate hypoxia (SpO_2 <94%) or hypotension (drop in SBP >20%)
- Grade 2: severe hypoxia (SpO_2 <88%) or hypotension (drop in SBP >40%) or altered mental status
- Grade 3: cardiovascular collapse requiring cardiopulmonary resuscitation

Fat embolism syndrome occurs more commonly than BCIS and should be considered in all patients with similar presentation. It manifests as the triad of respiratory failure, neurological impairment, and petechial rash (Bergman triad). Other findings may include the presence of fever, bradycardia, jaundice, the presence of fat particles in the blood, renal insufficiency, and retinal involvement. It often occurs within 24–48 h after surgery.

No specific treatment for BCIS exists and management is generally supportive. The systemic and pulmonary hemodynamic changes of right heart failure due to the elevated pulmonary arterial pressure and pulmonary vascular resistance, and reduced cardiac output secondary to bulging of the interventricular septum into the left ventricle, are transient and resolve with adequate fluid resuscitation and vasopressor therapy. In light of right heart failure, the use of sympathetic α_1-agonist vasopressor is recommended. The altered mental status that may occur is believed to be a result of femoral content embolization through a patent foramen ovale. Among the recommended measures to reduce the risk of BCIS include placing a venting hole in the femur and the use of noncemented prosthetics in high-risk patients.

TAKEAWAY POINTS

- BCIS is characterized by altered sensorium, hypoxia, hypotension, and cardiac arrhythmias after orthopedic cementation procedure.
- Fat embolism syndrome should be considered in the differential with BCIS.

FURTHER READING

Donaldson, A.J., Thomson, H.E., Harper, N.J., Kenny, N.W., 2009. Bone cement implantation syndrome. Br. J. Anaesth. 102 (1), 12–22.

Parisi, D.M., Koval, K., Egol, K., 2002. Fat embolism syndrome. Am. J. Orthop. 31 (9), 507–512.

Powell, J.N., McGrath, P.J., Lahiri, S.K., Hill, P., 1970. Cardiac arrest associated with bone cement. BMJ 3 (5718), 326.

Case 35

A 54-year-old male with history of hypothyroidism presents to the hospital complaining of cough, shortness of breath on exertion, weight loss, and fevers that have been progressively worsening over the last 2 weeks. A chest computed tomography is performed (Fig. 35.1) and he is started on antibiotics for suspected pneumonia. Over the next couple of days, his symptoms fail to improve and he undergoes a bronchoscopy with transbronchial biopsies. His cultures come back negative but his lung biopsy show positive IgG4 and CD138 staining (Fig. 35.2) of plasma cells on immunohistochemistry. Serum IgG4 levels were found to be elevated at 424 mg/dL.

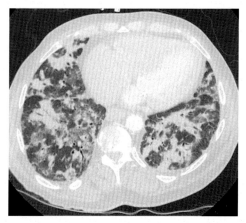

FIGURE 35.1   Chest computed tomography showing bilateral diffuse ground-glass opacities and patchy infiltrates.

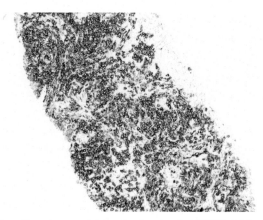

FIGURE 35.2   Lung tissue immunostaining showing abundant CD138 staining of plasma cells.

What is the diagnosis?

IgG4-RELATED SYSTEMIC DISEASE

IgG4-related systemic disease, also referred to as IgG4-related sclerosing disease, is a systemic fibroinflammatory disorder associated with elevated IgG4 levels. The first reported descriptions of elevated IgG4 levels involved patients with pancreatic manifestations, thereafter giving rise to the term systemic IgG4-related disease. It has also been associated with multiple organ involvement, including the lungs and mediastinum.

The disease is more commonly described in males, usually between the ages 55 and 65 years. The etiology of IgG4-realted disease remains unclear, but there is increasing evidence that an autoimmune mechanism involving naïve regulatory T-cells may be a factor. Whether the IgG4 antibodies are responsible for the pathogenesis of the disease or are the result of an inflammatory response remains to be determined.

Pulmonary involvement occurs in approximately 12–50% of cases, with hilar and mediastinal adenopathy being the most common manifestation of thoracic involvement, and nodular parenchymal lesions and bronchovascular bundle thickening being the most common pulmonary manifestations. Other pulmonary manifestations include ground-glass opacities, pulmonary infiltrates, bronchiectasis, pulmonary reticulations, tracheobronchial stenosis, lung entrapment, fibrosing mediastinitis, pleural effusions, pleural nodules, and interstitial lung disease.

Patients may be asymptomatic or present with nonspecific symptoms of cough, shortness of breath, hemoptysis, and chest pain. Constitutional symptoms of fevers, weight loss may occur but are uncommon. The diagnosis is based on the finding of elevated serum IgG4 levels (>135 mg/dL) that are present in 70–90% of cases. However, elevated IgG4 levels may also present in 5% of the normal population. On histology, there is lymphoplasmacytic tissue organ infiltration with IgG4-positive plasma cells that stain positive for CD138, and an elevated ratio of IgG4 to IgG positive plasma cells on immunohistochemical analysis. There is accompanying inflammation, obliterative phlebitis or arteritis, and fibrosis that have a characteristic storiform pattern. This storiform or "whorled" fibrosis pattern is often minimal or absent in the lungs where a collagenized fibrosis pattern is often seen with active fibroblastic proliferation. Eosinophilic infiltration may also be present.

Some cases of IgG4-related disease have been reported to resolve spontaneously; however, in most cases patients will require treatment with corticosteroids. In occasions where surgical resection of a focal lung lesion is undertaken, follow-up treatment with corticosteroids is usually not required. Corticosteroid sparing agents such as rituximab have been reported with clear clinical improvement. Increased risk of malignancy in these patients has been raised due to association with lymphoma, lung cancer, and pancreatic cancer, but this is yet to be determined.

TAKEAWAY POINTS

- IgG4-related systemic disease is a fibroinflammatory disorder of unclear etiology that may involve any organ.
- Patients with pulmonary involvement may present with nonspecific symptoms and various lung manifestations. Diagnosis is established on tissue biopsy and elevated IgG4 serum levels.

FURTHER READING

Hamano, H., Kawa, S., Horiuchi, A., Unno, H., Furuya, N., Akamatsu, T., et al., 2001. High serum IgG4 concentrations in patients with sclerosing pancreatitis. N. Engl. J. Med. 344 (10), 732–738.

Inoue, D., Zen, Y., Abo, H., Gabata, T., Demachi, H., Kobayashi, T., et al., 2009. Immunoglobulin G4-related lung disease: CT findings with pathologic correlations. Radiology 251 (1), 260–270.

Kamisawa, T., Funata, N., Hayashi, Y., Eishi, Y., Koike, M., Tsuruta, K., et al., 2003. A new clinicopathological entity of IgG4-related autoimmune disease. J. Gastroenterol. 38 (10), 982–984.

Ryu, J.H., Sekiguchi, H., Yi, E.S., 2012. Pulmonary manifestations of immunoglobulin G4-related sclerosing disease. Eur. Respir. J. 39 (1), 180–186.

Zen, Y., Inoue, D., Kitao, A., Onodera, M., Abo, H., Miyayama, S., et al., 2009. IgG4-related lung and pleural disease: a clinicopathologic study of 21 cases. Am. J. Surg. Pathol. 33 (12), 1886–1893.

Case 36

A 38-year-old male is evaluated for abnormal chest computed tomography (CT) findings. He is a nonsmoker with no relevant past medical history and no significant environmental exposures. His chest CT (Fig. 36.1) shows findings of bilateral infiltrates as well as apical pleural thickening. His pulmonary function tests (PFTs) are consistent with an intrinsic restrictive ventilatory defect with an elevated residual volume/total lung capacity (RV/TLC) ratio. Autoimmune serology testing is unrevealing. An open lung biopsy of the upper lobe is performed showing subpleural fibrosis with prominent elastosis (Figs. 36.2 and 36.3).

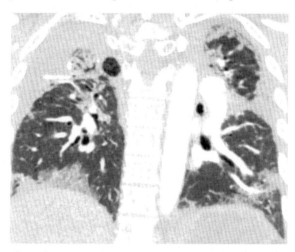

FIGURE 36.1 Coronal chest computed tomography showing bilateral parenchymal infiltrates and apical pleural thickening. *Courtesy of Tan-Lucien Mohammed, MD.*

FIGURE 36.2 Biopsy revealing dense subpleural fibrosis. *Courtesy of Mitra Mehrad, MD.*

FIGURE 36.3 Prominent elastosis with intervening collagen deposition. *Courtesy of Mitra Mehrad, MD.*

What is the diagnosis?

IDIOPATHIC PLEUROPARENCHYMAL FIBROELASTOSIS

Pleuroparenchymal fibroelastosis (PPFE) is a form of idiopathic interstitial pneumonia characterized by upper lobe predominant pleural and parenchymal fibrosis; it was first reported by Frankel in 2004. It may be related to a previously described entity, described in Japanese patients, termed pulmonary upper lobe fibrosis, or Amitani disease.

The exact etiology of the disease remains unknown, but it may be related to recurrent lower respiratory tract infections leading to elastin overproduction or impaired degradation. An association with smoking has not been noted. PPFE is characterized as being either idiopathic or secondary to various other conditions or exposures including ankylosing spondylitis, ulcerative colitis, malignancies, infections, gastrointestinal reflux disease, post bone marrow or lung transplantation, and exposure to asbestosis, radiation, or chemotherapy agents.

The disease is more common in males, presenting in the third to fifth decade of life. Patients may complain of slowly progressive shortness of breath on exertion, a nonproductive cough, and weight loss. They are prone to pneumothoraces (30%) and may present with acute onset chest pain. Idiopathic and secondary PPFE have been associated with the presence of a slender stature and a flattened thoracic cage on examination. It is unclear if this is a result of a congenital predisposition or secondary to the underlying disease process. Auscultation of the chest may reveal upper lobe inspiratory crackles in up to half of patients. Clubbing is not commonly noted.

PFTs will reveal an intrinsic restrictive ventilatory defect; however, due to compensatory hyperinflation of the middle and lower lobes, patients with IPPFE may have an increased RV to TLC ratio. Chest imaging shows bilateral apical and upper lobe pleural thickening with associated subpleural reticulation and nodules. The hilar structures may be elevated due to upper lobe volume loss. In advanced disease, these parenchymal findings may progress to fibrotic disease with architectural distortion, honeycombing, and bullae formation.

Diagnosis is established on lung biopsy. Histologically, there is evidence of intraalveolar and visceral pleura fibrosis with prominent subpleural and septal elastin fiber proliferation (elastosis). Parenchymal lymphocytic infiltration and fibroblastic foci may be present.

Treatment with steroids and other immunosuppressive agents have not been successful. As the disease progresses, the only therapeutic options are supportive care and referral to for lung transplant evaluation.

TAKEAWAY POINTS

- PPFE is a rare fibrotic lung disease that can be either idiopathic or secondary to other conditions/exposures.
- There is bilateral apical pleural thickening with subpleural fibrosis evidence on imaging. Lung biopsy with elastin staining can confirm the diagnosis.

FURTHER READING

Amitani, R., Niimi, A., Kuse, F., 1992. Idiopathic pulmonary upper lobe fibrosis (IPUF). Kokyu 11, 693–699.

Frankel, S.K., Cool, C.D., Lynch, D.A., Brown, K.K., 2004. Idiopathic pleuroparenchymal fibro-elastosis: description of a novel clinicopathologic entity. Chest 126 (6), 2007–2013.

Reddy, T.L., Tominaga, M., Hansell, D.M., von der Thusen, J., Rassl, D., Parfrey, H., et al., 2012. Pleuroparenchymal fibroelastosis: a spectrum of histopathological and imaging phenotypes. Eur. Respir J. 40 (2), 377–385.

Case 37

A 28-year-old male with history of celiac disease presents with fevers and short-ness of breath. A chest X-ray shows bilateral infiltrates and his chest computed tomography is concerning for active alveolar bleeding (Fig. 37.1). He undergoes a bronchoscopy with serial lavages that is consistent with diffuse alveolar hemor-rhage with bronchoalveolar lavage (BAL) showing abundant hemosiderin-laden macrophages. Transbronchial biopsies do not show any evidence of vasculitis or capillaritis. Serum antineutrophil cytoplasmic antibodies, antiglomerular base-ment membrane, and other autoimmune antibodies are all negative.

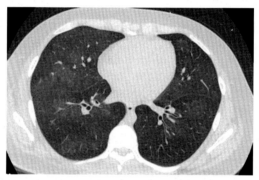

FIGURE 37.1   Chest computed tomography revealing patchy bilateral peribronchial ground-glass opacities in the middle and lower lobes concerning for alveolar hemorrhage.

What is the diagnosis?
What other conditions have been associated with this disease?

IDIOPATHIC PULMONARY HEMOSIDEROSIS

Idiopathic pulmonary hemosiderosis (IPH) is characterized by recurrent episodes of diffuse alveolar hemorrhage resulting in alveolar hemosiderosis in the absence of vasculitis, capillaritis, or any other etiology. The free iron accumulates in the alveoli leading to increased free radical formation and subsequent pulmonary fibrosis. The first description of IPH was made by Virchow in 1864 in a patient who was described as having "brown lung induration." The disease is more common in the pediatric population but 20% of cases occur in adults.

The etiology of IPH remains unclear; however, the immune system seems to play a role in disease pathogenesis since patients with IPH respond to immunosuppressive therapy. IPH has been associated with celiac disease (Lane-Hamilton syndrome) and immunoglobulin A gammopathy.

IPH is more commonly reported in males. Patients may present with fevers, shortness of breath, coughing, and hemoptysis. Laboratory analysis will reveal iron deficiency anemia. Pulmonary function tests will be consistent with restrictive lung disease with evolving fibrosis. Chest imaging during episode of acute hemorrhage may demonstrate bilateral ground-glass opacities in the middle and lower lobes. BAL will be positive for hemosiderin-laden macrophages. Diagnosis requires exclusion of other possible etiologies and a lung biopsy negative for capillaritis, granulomas, or other pathology (bland hemorrhage).

Prognosis is poor with patients developing fibrosis within 5 years of initial presentation. Episodes of acute hemorrhage may resolve spontaneously but seem to respond to high-dose corticosteroids. Patients with severe acute hemorrhage and recurrent episodes may benefit from maintenance therapy with steroid sparing immunosuppressive treatments. In patients with associated celiac disease, a gluten-free diet has been to linked to disease remission. Reports of hemorrhage recurrence after lung transplantation have been reported.

TAKEAWAY POINTS

- IPH should be suspected in individuals with iron deficiency anemia and recurrent episodes of pulmonary hemorrhage in the absence of any obvious etiology.
- Diagnosis requires the presence of hemosiderin-laden macrophages and absence of alveolar capillaritis on lung biopsy.
- Patients with Lane-Hamilton syndrome respond to a gluten-free diet.

FURTHER READING

Ioachimescu, O.C., Sieber, S., Kotch, A., 2004. Idiopathic pulmonary haemosiderosis revisited. Eur. Respir. J. 24 (1), 162–170.

Milman, N., Pedersen, F.M., 1998. Idiopathic pulmonary haemosiderosis. Epidemiology, pathogenic aspects and diagnosis. Respir. Med. 92 (7), 902–907.

Virchow, R., 1864. Die Krankhaften Geshwulste. August Hirschwald, Berlin, pp. 26–240.

Case 38

A 32-year-old Asian female is being evaluated for a 10-day history of fever. Her infectious and autoimmune workup has been unrevealing and she undergoes a computed tomography (CT) scan of the head, chest, abdomen, and pelvis that reveals moderately enlarged posterior cervical and mediastinal lymph nodes. She undergoes a biopsy of her cervical lymph node that is negative for malignancy and infectious etiology but reveals evidence of paracortical necrosis, crescentic histiocytes infiltration, and karyorrhectic debris.

What is the diagnosis?
What is the treatment for this disease?

KIKUCHI–FUJIMOTO DISEASE

Kikuchi–Fujimoto disease (KFD), also known as histiocytic necrotizing lymphadenitis, is a benign, self-limited form of lymphadenitis that was first described by Kikuchi in Japan in 1972.

The etiology of the disease remains unclear; however, it has been postulated that KFD may be a form of autoimmune disease resulting from a viral infection, since patients frequently have a flu-like prodrome followed by a self-limited course. The lymphadenopathy occurs as a result of CD8+ T-cell mediated apoptotic cell death.

KFD commonly affects young females of Asian descent, who present with moderately enlarged (1–2 cm) cervical, nontender, lymphadenopathy. The posterior cervical lymph nodes are most frequently involved but patients may have adenopathy at other sites such as the axilla, inguinal area, mediastinum, or mesentery. Systemic manifestations including fevers, night sweats, fatigue, and arthralgias may be present. Laboratory testing may show an elevated erythrocyte sedimentation rate and half of patients will have leukopenia. KFD often occurs in association with systemic lupus erythematosus (SLE) and other autoimmune diseases, with reports of SLE developing years after the lymphadenopathy has resolved.

Pulmonary involvement, while uncommon, may include:

- Mediastinal and hilar adenopathy
- Parenchymal infiltrates
- Pleural effusions
- Cryptogenic organizing pneumonia
- Pulmonary hemorrhage
- Respiratory failure

It is important to distinguish KFD from lymphoma, tuberculosis, sarcoidosis, and other infectious and autoimmune conditions. CT imaging will reveal homogenous lymph node enlargement with areas of focal necrosis. The diagnosis is established on lymph node biopsy with histology revealing paracortical foci of necrosis, crescentic histiocytes proliferation, and karyorrhectic debris (histiocytes and macrophages containing phagocytized debris).

KFD is often self-limited, with recovery period from 1 to 4 months. It rarely recurs. Treatment is supportive with analgesics and antipyretics. Corticosteroids are recommended for patients with significant extranodal disease.

TAKEAWAY POINTS

- KFD is a benign, self-limiting cause of cervical lymphadenopathy in young females, with infrequent pleuroparenchymal manifestations.
- Patients should be followed after disease resolution for the development of SLE.

FURTHER READING

Garcia-Zamalloa, A., Taboada-Gomez, J., Bernardo-Galan, P., Magdalena, F.M., Zaldumbide-Duenas, L., Ugarte-Maiztegui, M., 2010. Bilateral pleural effusion and interstitial lung disease as unusual manifestations of Kikuchi–Fujimoto disease: case report and literature review. BMC Pulm. Med. 10, 54.

Kikuchi, M., 1972. Lymphadenitis showing focal reticulum cell hyperplasia with nuclear debris and phagocytes. Acta Hematol. Jpn. 35, 379–380.

Kucukardali, Y., Solmazgul, E., Kunter, E., Oncul, O., Yildirim, S., Kaplan, M., 2007. Kikuchi–Fujimoto disease: analysis of 244 cases. Clin. Rheumatol. 26 (1), 50–54.

Naito, N., Shinohara, T., Machida, H., Hino, H., Naruse, K., Ogushi, F., 2015. Kikuchi–Fujimoto disease associated with community acquired pneumonia showing intrathoracic lymphadenopathy without cervical lesions. SpringerPlus 4, 693.

Case 39

A 32-year-old female with a history of recurrent pneumothorax is evaluated for shortness of breath on exertion on climbing stairs. She is a nonsmoker, with no family history of lung diseases, and no significant exposures. Her physical examination was unremarkable. Pulmonary function testing reveals a mild obstructive ventilatory defect, and her chest X-ray shows a diffuse interstitial pattern with preserved lung volumes. Her chest computed tomography (CT) scan is shown in Fig. 39.1.

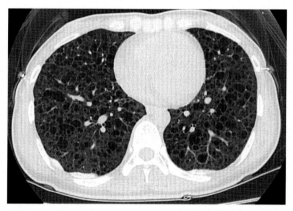

FIGURE 39.1   Chest computed tomography scan showing diffuse, bilateral, well-circumscribed, thin-walled cysts.

What is the diagnosis?

LYMPHANGIOLEIOMYOMATOSIS

Lymphangioleiomyomatosis (LAM) is a rare progressive disease that predominately affects premenopausal women of childbearing age. It was first described by Van Stossel in 1937 as muscular cirrhosis of the lungs. Almost one-third of cases occur in women with tuberous sclerosis complex (TSC), while the majority are sporadic, occurring in the absence of any TSC manifestations.

LAM is thought to result from mutations in the TSC genes (TSC1 and TSC2), resulting in the activation of the mechanistic target of rapamycin (mTOR) signaling pathway, leading to the abnormal diffuse hamartomatous proliferation of smooth muscle and spindle cells (LAM cells) in the lung parenchyma, airways, blood vessels, and lymphatic vessels. Sporadic LAM is associated with TSC2 gene mutations while TSC-LAM may occur with either gene mutation.

Pulmonary manifestations include cystic lung disease, recurrent pneumothoraces, and chylous pleural effusions. Extrapulmonary manifestations may include retroperitoneal adenopathy (77%), angiomyolipomas (50%), chylous ascites, lymphedema, and osteoporosis. The disease primarily affects the lungs with pulmonary symptoms often being the first manifestation of disease. Patients may complain of shortness of breath, coughing, chest pain, or hemoptysis. Patients may rarely present with symptoms associated with extrapulmonary disease, such as bleeding into a renal angiolipoma. Pulmonary symptoms may be exacerbated by exogenous estrogen, menstruation, and pregnancy.

Peribronchovascular lymphatic dilation may lead to airway narrowing and obstruction. Pulmonary function testing often reveals an obstructive ventilatory defect with increased lung volume and reduced DLCO; however, normal, restrictive, and mixed patterns may also occur. Chest CT shows innumerable, bilateral, thin-walled parenchymal cysts without lung nodules. There is often preservation of lung volume despite the parenchymal involvement. Pleural effusions occur in one-third of cases, and are often chylous, unilateral, and lymphocyte predominant. In cases of nondiagnostic CT findings, tissue biopsy will reveal abnormal proliferation of smooth muscle cells that stain positive with the monoclonal antibody HMB-45, smooth muscle actin, myosin, and desmin. Elevated serum levels of vascular endothelial growth factor D (>800 pg/mL) are often found in patients with sporadic LAM, and can assist in making the diagnosis in the setting of typical CT findings.

Management is directed at treating different manifestations of the disease. Sirolimus (Rapamycin) has been shown to slow lung function decline and reduce the size of angiolipomas in LAM by inhibiting activation of the mTOR pathway. Renal angiomyolipomas are often surgically resected. In progressive cases, lung transplantation should be considered; however, the disease has been reported to recur after transplant.

TUBEROUS SCLEROSIS COMPLEX

TSC is an autosomal dominant disorder, characterized by benign tumors, seizures, mild cognitive impairment, and pulmonary LAM in up to one-third of cases. On examination, patients may have skin manifestations of facial angiofibromas, ash-leaf spots, and shagreen patches.

Other lymphatic disorders that may affect the lungs include:

- Pulmonary lymphangiectasis: abnormal dilation of lymphatics that may be either congenital (usually fatal in infants) or secondary (seen in children).
- Lymphangiomas: focal lymphatic tissue proliferation resulting in multicystic structures (e.g., cystic hygromas)
- Lymphangiomatosis: abnormal proliferation of lymphatic vessels in different organs that presents early in life. Thoracic lymphangiomatosis can present as pulmonary or mediastinal masses, and chylous effusions. Bone lesions are common.
- Lymphatic dysplasia syndrome: consists of various syndromes without an identifiable cause and includes idiopathic lymphedema syndromes, yellow nail syndrome, congenital chylothorax, and idiopathic chylous effusions.

TAKEAWAY POINTS

- LAM is the abnormal proliferation of smooth muscle cells manifesting as diffuse cystic lung diseases in females of childbearing age. Lung volumes are increased or preserved despite lung involvement.
- The majority of patients with LAM are sporadic. Over one-third of cases are associated with TSC.
- Sirolimus has been shown to slow pulmonary function decline and decrease the size of angiomyolipomas in patients with LAM.

FURTHER READING

Van Stossel, E., 1937. Ueber muskulare cirrhose der lunge. Beiträge zur Klinik der Tuberkulose und spezifischen Tuberkulose-Forschung 90, 432–442.

Taylor, J.R., Ryu, J., Colby, T.V., Raffin, T.A., 1990. Lymphangioleiomyomatosis. Clinical course in 32 patients. N. Engl. J. Med. 323 (18), 1254–1260.

Faul, J.L., Berry, G.J., Colby, T.V., Ruoss, S.J., Walter, M.B., Rosen, G.D., et al., 2000. Thoracic lymphangiomas, lymphangiectasis, lymphangiomatosis, and lymphatic dysplasia syndrome. Am. J. Respir. Crit. Care Med. 161 (3 Pt 1), 1037–1046.

McCormack, F.X., Inoue, Y., Moss, J., Singer, L.G., Strange, C., Nakata, K., et al., 2011. Efficacy and safety of sirolimus in lymphangioleiomyomatosis. N. Engl. J. Med. 364 (17), 1595–1606.

Lama, A., Ferreiro, L., Golpe, A., Gude, F., Alvarez-Dobano, J.M., Gonzalez-Barcala, F.J., et al., 2016. Characteristics of patients with lymphangioleiomyomatosis and pleural effusion: a systematic review. Respiration 91 (3), 256–264.

Case 40

A 57-year-old male with history of multiple myeloma undergoing evaluation for a bone marrow transplant is referred for evaluation of abnormal chest computed tomography (CT) findings (Fig. 40.1). He has no history of smoking and does not endorse any pulmonary symptoms. He undergoes bronchoscopy and transbronchial biopsies. Histological examination reveals abundant extracellular amorphous deposits in the lung parenchyma with negative apple-green birefringence with Congo red stains.

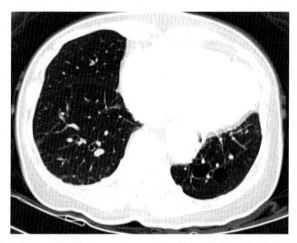

FIGURE 40.1   Chest computed tomography revealing thin-walled cysts in the left lower lobe.

What is the diagnosis?

PULMONARY LIGHT CHAIN DEPOSITION DISEASE

Light chain deposition disease (LCDD) was first described by Randall in 1976 and is characterized by the systemic deposition of nonamyloid amorphous material composed of monotypic immunoglobulin light chains secreted by a clone of plasma cells. The light chains in LCDD are almost always composed of kappa (κ) light chains; however, lambda (λ) light chain deposits may also occur. LCDD commonly occurs in the setting of a lymphoproliferative disorder, with multiple myeloma accounting for around 75% of cases, but may be seen with autoimmune diseases (Sjogren disease), or be idiopathic.

The kidneys, heart, and liver are the most frequently involved organs, while isolated pulmonary involvement is rare. Patients with pulmonary LCDD often present in their 5th or 6th decade complaining of dyspnea on exertion or a nonproductive cough. Pulmonary involvement is associated most frequently with the presence of diffuse or localized thin-walled cysts, with or without the presence pulmonary nodules and small airway dilation on CT imaging. The lung cysts often have traversing blood vessels and with no preferential zone distribution. The cyst formation in LCDD is a result of elastolysis by metalloproteases, a mechanism similar to that seen with Langerhans cell histiocytosis and lymphangioleiomyomatosis.

The diagnosis is established based on histopathological findings and features that allow it to be distinguished from pulmonary amyloidosis. Biopsy will show eosinophilic granular amorphous deposits that stain positively for kappa light chains, on a background of lymphocytic infiltrates within the alveolar walls, small airways, and vessels. The amorphous material lacks the typical fibrillary structure, β-pleated configuration, and Congo red stain features typical of amyloidosis.

Pulmonary LCDD is a progressive disease with a poor prognosis leading to respiratory failure overtime. Treatment is directed at the underlying lymphoproliferative disorder, when present, with cytotoxic chemotherapy or stem cell transplantation. In advanced pulmonary cases, patients may require lung transplantation. All patients with pulmonary LCDD should be assessed for possible renal (proteinuria, microscopic hematuria, renal failure) and cardiac (arrhythmias, diastolic dysfunction, heart failure) involvement.

TAKEAWAY POINTS

- Pulmonary LCDD is a light chain deposition disease that shares many features with amyloidosis but is distinguishable from it based on histopathological findings.
- Patients often have cystic lung lesions with or without pulmonary nodules, a CT pattern that may resemble Langerhans cell histiocytosis, lymphangioleiomyomatosis, and lymphoid interstitial pneumonia.

FURTHER READING

Colombat, M., Stern, M., Groussard, O., Droz, D., Brauner, M., Valeyre, D., et al., 2006. Pulmonary cystic disorder related to light chain deposition disease. Am. J. Respir. Crit. Care Med. 173 (7), 777–780.

Randall, R.E., Williamson Jr., W.C., Mullinax, F., Tung, M.Y., Still, W.J., 1976. Manifestations of systemic light chain deposition. Am. J. Med. 60 (2), 293–299.

Seaman, D.M., Meyer, C.A., Gilman, M.D., McCormack, F.X., 2011. Diffuse cystic lung disease at high-resolution CT. Am. J. Roentgenol. 196 (6), 1305–1311.

Sheard, S., Nicholson, A.G., Edmunds, L., Wotherspoon, A.C., Hansell, D.M., 2015. Pulmonary light-chain deposition disease: CT and pathology findings in nine patients. Clin. Radiol. 70 (5), 515–522.

Case 41

A 57-year-old male with a history of severe chronic obstructive pulmonary disease undergoes single left lung transplantation without any complications. His immunosuppressive maintenance regimen consists of prednisone, mycophenolate mofetil, and azathioprine. Three weeks after his transplant, he presents with a 3-day history of fevers, fatigue, diarrhea, and a pruritic maculopapular rash on his arms and legs.

Surveillance bronchoscopy with transbronchial lung biopsies done a few days prior to symptom onset did not show any evidence of rejection. Laboratory analysis reveals new onset pancytopenia and transaminitis (all previously normal). A thorough workup does not reveal any evidence of infection. A skin biopsy is performed and reveals dermal perivascular lymphocytic infiltration with apoptotic cells. Based on these findings, a polymerase chain reaction (PCR) based analysis of the peripheral blood is done and shows that >90% of lymphocytes are of donor origin.

What is the diagnosis?

GRAFT-VERSUS-HOST DISEASE AFTER LUNG TRANSPLANTATION

Graft-versus-host disease (GVHD) describes the engraftment of transfused immunocompetent donor lymphocytes into an immunosuppressed host leading to an immunologically mediated attack against the host tissues. It is commonly reported after hematopoietic stem cell transplantation, but has also been described following transfusion of blood products and after solid organ transplantation, such as liver and small bowel. Only a few cases of GVHD after lung transplant have been described; however, this may be related to underrecognition of the disease.

GVHD affects the skin, gastrointestinal tract, hepatobiliary system, and bone marrow resulting in the clinical manifestations of pruritic maculopapular skin rash, diarrhea, elevated liver enzymes, fevers, and pancytopenia, a few weeks after transplantation. Symptoms may often be mistakenly attributed to side effects of immunosuppressive medication or an infection.

Diagnosis is established based on clinical suspicion and biopsy (usually of the skin) demonstrating donor lymphocytes chimerism in the host. Host peripheral blood or bone marrow chimerism analysis using PCR or fluorescence in situ hybridization analysis to demonstrate donor DNA or sex chromosomal differences (if relevant), may also be used to confirm the diagnosis.

Prognosis is poor, but early recognition may improve outcomes. High-dose corticosteroids are usually given, frequently along with other immunosuppressive agents including calcineurin inhibitors, mycophenolate mofetil, antithymocyte globulin, or monoclonal antibodies (anti-IL-2 receptors). Mortality is high and is frequently related to infectious complications, which result from escalated immunosuppressive therapy.

TAKEAWAY POINTS

- GVHD after lung transplantation is rare and manifests as fevers, gastrointestinal complaints, skin rashes, elevated liver enzymes, and pancytopenia.
- Diagnosis is established on tissue biopsy with evidence of donor lymphocyte chimerism within the host.
- Mortality is high despite corticosteroids and other immunosuppressive agents, likely related to delayed recognition.

FURTHER READING

Fossi, A., Voltolini, L., Filippi, R., et al., 2009. Severe acute graft versus host disease after lung transplant: report of a case successfully treated with high dose corticosteroids. J. Heart Lung Transplant. 28, 508–510.

Luckraz, H., Zagolin, M., McNeil, K., Wallwork, J., 2003. Graft-versus-host disease in lung transplantation: 4 case reports and literature review. J. Heart Lung Transplant. 22, 691–697.

Worel, N., Bojic, A., Binder, M., et al., 2008. Catastrophic graft-versus-host disease after lung transplantation proven by PCR-based chimerism analysis. Transpl. Int. 21, 1098–1101.

Case 42

A 78-year-old male presents with a history of progressive dyspnea on exertion and a nonproductive cough. He has been treated for recurrent pneumonias in the lower lobes over the last few years. He has a history of coronary artery disease, stroke with residual deficit and functional dysphagia, diabetes mellitus, and chronic constipation for which he uses mineral oil. A chest X-ray shows infiltrates in the lower lobes and he is treated with antibiotics for presumed pneumonia. He fails to improve with therapy so he undergoes a chest computed tomography (CT) (Fig. 42.1) and bronchoscopy with bronchoalveolar lavage (BAL). BAL cytology shows evidence of many foamy macrophages with large cytoplasmic vacuoles that stain positive with Sudan III stain.

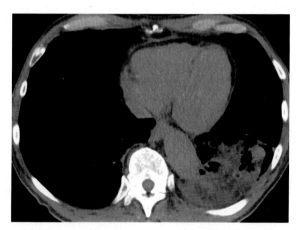

FIGURE 42.1   Chest computed tomography scan showing a fat density left lower lobe consolidation. *Courtesy of Tan-Lucien Mohammed, MD.*

What is the diagnosis?
What are the different etiologies of this disease?

LIPOID PNEUMONIA

Lipoid pneumonia is the accumulation of lipid material within the pulmonary alveoli as a result of exogenous or endogenous etiologies. It was first described by Laughlen in 1925 in patients with a history of use of chronic laxatives or oil-based nasal drops. It has also been referred to as cholesterol or golden pneumonia.

Exogenous lipoid pneumonia may be acute or chronic in onset depending on the etiology and extent of exposure. Acute lipoid pneumonia has been reported as the result of accidental poisoning due to aspiration or inhalation of petroleum-based products (gas siphoning) as well as in "fire-eater" performers who may inhale the petroleum-based fluid kerdan as part of their act. Chronic lipoid pneumonia on the other hand has been reported in individuals with a history of prolonged use of fat-containing products or oils such as laxatives, mineral oils, lip balm, and nose drops, often in the setting of an anatomical or functional swallowing impairment.

Endogenous lipoid pneumonia occurs as a result of lipid accumulation within the alveoli and intraalveolar macrophages. It has been reported in the setting of a postobstructive process (lung cancer), pulmonary alveolar proteinosis, chronic pulmonary infections, or a lipid storage disorder (Neiman-Pick syndrome).

Patients with acute onset of lipoid pneumonia may present with acute onset cough, dyspnea, and respiratory failure and may be misdiagnosed as having a community-acquired pneumonia. The chronic form of the disease has a more indolent presentation with patients being asymptomatic or presenting with gradually progressive cough, dyspnea on exertion, and low-grade fevers. Chest CT may reveal findings of nodules, consolidation, or ground-glass opacities in a peribronchovascular distribution, often within the middle or lower lobes. In some cases, patients may have findings of crazy-paving pattern on imaging. The nodules, masses, or consolidations may contain areas with low attenuation (<-30 Hounsfield units), consistent with the presence of fat.

Diagnosis may also be established on BAL or biopsy, showing findings of lipid vacuoles within macrophages (lipid-laden or foamy macrophages) in the lavage fluid or alveoli (Fig. 42.2). The finding of large cytoplasmic vacuoles within the foamy macrophages may help distinguish exogenous lipoid pneumonia from the endogenous type that may have small cytoplasmic vacuoles. The use of a specific fat stain (Sudan III or IV or Oil Red O) on pathology specimens will confirm the presence of lipids.

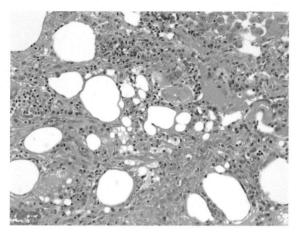

FIGURE 42.2   Lung histology showing many histiocytes and multinucleated giant cells surrounding lipid vacuoles, with evidence of chronic inflammation. *Courtesy of Humberto Trejo Bittar, MD.*

Treatment for the lipoid pneumonia is supportive or may be directed at the underlying etiology (chronic aspiration, infection, pulmonary alveolar proteinosis). Patients should be educated to avoid exposure to potential inciting agents, and aspiration risk should be assessed and minimized. In cases of chronic repetitive aspiration, lipoid pneumonia may progress to pulmonary fibrosis.

TAKEAWAY POINTS

- Lipoid pneumonia may be exogenous or endogenous in nature. A thorough history may be required to reveal inciting exogenous cause.
- Areas of negative attenuation within pulmonary lesions on chest CT are consistent with fat and can aid in making the diagnosis.

FURTHER READING

Laughlen, G.F., 1925. Studies on pneumonia following naso-pharyngeal injections of oil. Am. J. Pathol. 1 (4), 407–414.1.

Marchiori, E., Zanetti, G., Mano, C.M., Hochhegger, B., 2011. Exogenous lipoid pneumonia. Clinical and radiological manifestations. Respir. Med. 105 (5), 659–666.

Spickard 3rd, A., Hirschmann, J.V., 1994. Exogenous lipoid pneumonia. Arch. Intern. Med. 154 (6), 686–692.

Case 43

A 49-year-old male presents with complains of a nonproductive cough, low-grade fevers, chills, and unintentional weight loss over a period of 6 months. He is a nonsmoker with a history of coronary artery disease and rheumatoid arthritis. He denies any previous exposure to tuberculosis. His physical examination is unremarkable. Chest computed tomography (CT) scan is shown in Fig. 43.1. His sputum cultures and acid-fast bacilli stains are negative and antineutrophil cytoplasmic antibodies testing is negative.

Patient undergoes transbronchial biopsies that reveal a mononuclear cell infiltrate of large atypical CD20+ B cells that infiltrate the blood vessel walls, with a background of small benign CD3+ T cells. There is an area of central necrosis and the in situ hybridization is positive for Epstein-Barr virus (EBV) RNA (Fig. 43.2).

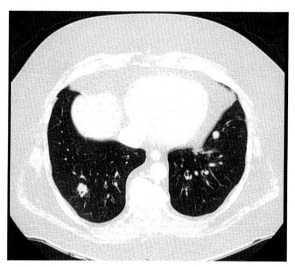

FIGURE 43.1   Chest computed tomography scan showing bilateral pulmonary nodules.

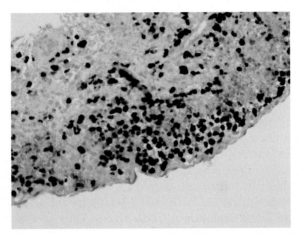

FIGURE 43.2 Lung biopsy showing positive staining for Epstein-Barr virus RNA in the large atypical B cells.

What is the diagnosis?

PULMONARY LYMPHOMATOID GRANULOMATOSIS

Lymphomatoid granulomatosis (LYG), also known as angiocentric lymphoma, is an EBV driven lymphoproliferative disorder that was first described by Leibow in 1972 and is considered as part of the spectrum of EBV-associated B cell lymphomas.

The lungs are primarily involved in over 90% of cases, followed by skin (50%) and central nervous system (25%) involvement. LYG is twice as common in males and often manifests in the third to fifth decade. Patients with pulmonary LYG may present with a nonproductive cough, shortness of breath, chest pain, fevers, night sweats, weight loss, and occasionally an erythematous rash. Affected individuals frequently have risk factors associated with immune dysfunction that may predispose them to develop LYG. These include rheumatoid arthritis, Sjogren syndrome, organ transplant, HIV infection, chronic viral hepatitis, and use of immunosuppressive medications (azathioprine, methotrexate, and imatinib).

Chest CT imaging will most commonly have findings of bilateral poorly defined nodules along the peribronchovascular bundles throughout the lungs. These are frequently peripherally present in the lower lung fields and may be migratory in nature. Other described findings include pulmonary masses, consolidations, cavitary lesions, thin-walled cysts, and pleural effusions. Mediastinal adenopathy is uncommon and its presence should suggest another process (lymphoma, necrotizing sarcoidosis).

The diagnosis is established on tissue biopsy. Skin biopsy should be pursued first if a rash is present; otherwise lung biopsy should be performed, preferably via video-assisted thoracoscopic surgery. The classical histological triad consists of:

- Nodular polymorphic lymphoid infiltrates (atypical CD20+ B cells with a background of reactive CD3+ T cells)
- Transmural lymphoid infiltration of arteries and veins (angiitis)
- Focal areas of necrosis within the lymphoid nodules (granulomatosis)

WHO grades LYG according to the amount of atypical B cell present (Grades 1–3). EBV RNA, identified by in situ hybridization in the large B cells, is present in the majority of cases and supports the diagnosis of LYG.

LYG is a progressive disorder with an estimated survival of 63–90% at 5 years. Asymptomatic patients with low-grade disease are usually observed without treatment. There may be spontaneous remission in 20% of cases. Associated conditions should be treated and medication associated with development of LYG should be discontinued if possible. For symptomatic, high-grade and progressive disease with extrapulmonary involvement, treatment involves chemotherapy along with interferon-α-2b or rituximab.

TAKEAWAY POINTS

- Pulmonary LYG is an EBV-associated lymphoproliferative disorder that often presents with bilateral, migratory lung nodules in the lower lung fields. It is important to distinguish this from other vasculitic disorders and lymphomas.
- Patients with low-grade disease are observed for radiological change and may have spontaneous remission. High-grade or symptomatic disease will require chemotherapy treatment.

FURTHER READING

Colby, T.V., 2012. Current histological diagnosis of lymphomatoid granulomatosis. Mod. Pathol. 25 (Suppl. 1), S39–S42.

Katzenstein, A.L., Doxtader, E., Narendra, S., 2010. Lymphomatoid granulomatosis: insights gained over 4 decades. Am. J. Surg. Pathol. 34 (12), e35–e48.

Liebow, A.A., Carrington, C.R., Friedman, P.J., 1972. Lymphomatoid granulomatosis. Hum. Pathol. 3 (4), 457–558.

Case 44

A 46-year-old male smoker, with a history of recurrent respiratory tract infections, presents with a productive cough and fevers. He is diagnosed with a right lower lobe pneumonia and is started on antibiotic treatment. A chest X-ray (CXR) reveals an enlarged trachea measuring 3.6 cm. He undergoes a chest computed tomography (CT) scan that confirms enlargement of the trachea and central bronchi (Fig. 44.1) and posterior tracheal diverticuli (Fig. 44.2).

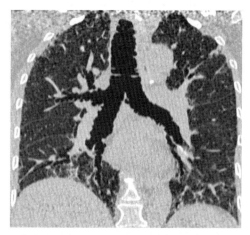

FIGURE 44.1 Coronal chest computed tomography scan showing tracheobronchomegaly with evidence of subpleural fibrosis.

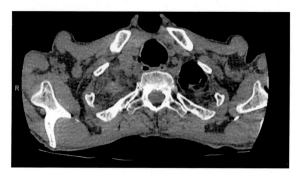

FIGURE 44.2 Chest computed tomography scan showing a dilated trachea and posterior tracheal diverticuli.

What is the diagnosis?

Rare and Interesting Cases in Pulmonary Medicine.

MOUNIER-KUHN SYNDROME

Mounier-Kuhn syndrome, also known as idiopathic tracheobronchomegaly, is a congenital disorder characterized by a dilated trachea and proximal bronchi; it was first described in 1932.

Mounier-Kuhn syndrome occurs as a result of absence or atrophy of the elastic fibers of the trachea and proximal bronchi, leading to thinning of the smooth muscle layers and resultant dilatation and tracheobronchomalacia. There is impaired airway clearance and pooling of secretions in the dilated airways leading to recurrent respiratory tract infections. The etiology remains unclear; however, the description of familial cases of Mounier-Kuhn syndrome has led to the suggestion that it may be an autosomal recessive inherited disorder. It has been associated with connective tissue disorders such as Ehler-Danlos syndrome, Marfans syndrome, and cutis laxa. Mounier-Kuhn syndrome has also been associated with absence of the myenteric plexus in the trachea wall.

There are three types of the disease:

- Type I: symmetrical diffuse enlargement of both trachea and bronchi.
- Type II: eccentric enlargement with pronounced diverticula and abrupt change to normal bronchial size (most common type).
- Type III: diverticula may extend to the more distal bronchi.

Mounier-Kuhn syndrome is more common in males, who often present in their 3rd or 4th decade with recurrent lower respiratory tract infections. They usually complain of dyspnea, a dry or productive cough, hemoptysis, and chronic rhinosinusitis. Over time they may develop bronchiectasis with associated cough productive of thick purulent secretions. Clubbing is a common finding among affected individuals.

Pulmonary function testing may be normal or show evidence of an obstructive ventilatory defect and an increased residual volume consistent with increased dead space. The CXR will reveal tracheobronchomegaly, although this is often missed or underappreciated. The chest CT will reveal an enlarged trachea and proximal bronchi with an abrupt transition to normal caliber airways distal to the dilatation. Diagnostic airway diameter sizes for Mounier-Kuhn syndrome in adults are a transverse tracheal diameter >30 mm measured 2 cm above the aortic arch, a right bronchial diameter >24 mm, and a left bronchial diameter >23 mm. There may be posterior protrusions of the musculomembranous tissue between the trachea folds resulting in saclike outpouchings. A dynamic chest CT will uncover variations in the airway diameter as they collapse on expiration. Bronchoscopy will show the diverticuli and help confirm the tracheobronchomegaly and airway collapse. The airway diameter may be seen to enlarge upon performing the Valsalva maneuver and decrease on doing the Muller maneuver.

Individuals may develop bronchiectasis, emphysema with bullous disease, and pulmonary fibrosis. Treatment is often supportive, aimed at managing airway clearance and treating respiratory infections. Some patients with tracheobronchomalacia may benefit from airway stenting in severe cases. There is a report of one successful lung transplant in a patient with this disease.

WILLIAMS-CAMPBELL SYNDROME

Williams-Campbell syndrome (WCS) is a congenital disorder of the bronchial cartilages characterized by atrophy or absence of the subsegmental bronchial cartilages of the fourth to sixth order bronchial divisions, leading to dilation of the distal bronchi. Mounier-Kuhn syndrome and WCS are distinct as the subsegmental bronchi are of normal caliber in Mounier-Kuhn syndrome, and patients with WCS have normal caliber trachea and proximal bronchi. In WCS, expiratory collapse of the dilated cystic bronchi may be appreciated on dynamic imaging of the chest. As in Mounier-Kuhn syndrome, patients with WCS are prone to develop recurrent lower respiratory tract infections because of difficulty clearing secretions from the dilated bronchi. Treatment is also supportive and includes use of airway clearance techniques and antibiotics for respiratory infections.

TAKEAWAY POINTS

- Patients with Mounier-Kuhn syndrome have dilated trachea and central bronchi, while patients with WCS have dilated subsegmental bronchi with sparing of the trachea and proximal bronchi.
- Patients are prone to developing recurrent respiratory infections and bronchiectasis. Treatment in both cases is supportive.

FURTHER READING

Krustins, E., Kravale, Z., Buls, A., 2013. Mounier-Kuhn syndrome or congenital tracheobroncho-megaly: a literature review. Respir. Med. 107 (12), 1822–1828.

Mounier-Kuhn, P., 1932. Dilatation de la trachée: constatations radiographiques et bronchoscopiques. Lyon Med. 150, 106–109.

Noriega Aldave, A.P., William Saliski, D., 2014. The clinical manifestations, diagnosis and management of Williams-Campbell syndrome. N. Am. J. Med. Sci. 6 (9), 429–432.

Williams, H., Campbell, P., 1960. Generalized bronchiectasis associated with deficiency of cartilage in the bronchial tree. Arch. Dis. Child. 35, 182–191.

Case 45

A 43-year-old Hispanic female is evaluated for slowly progressive shortness of breath on exertion for 10 years. She told that many years ago that she had an abnormal chest X-ray (CXR), but she did not pursue further workup. Her physical examination is remarkable for bilateral inspiratory crackles in all lung fields, and her pulmonary function tests (PFTs) results are as follows: forced expiratory volume 62%, forced vital capacity 59%, forced expiratory volume 1/forced vital capacity ratio 107 of predicted, total lung capacity 54%, and diffusion capacity of the lung for carbon monoxide 41%. A CXR and chest computed tomography (CT) are performed (Figs. 45.1 and 45.2).

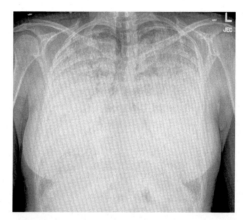

FIGURE 45.1   Chest X-ray showing diffuse fine sand-like micronodules in all lung fields.

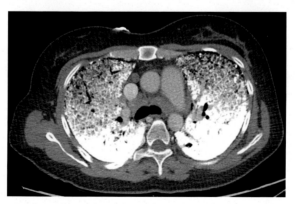

FIGURE 45.2   Chest computed tomography scan demonstrating diffuse calcified micronodules, more predominant in the subpleural and basal segments.

What is the diagnosis?

PULMONARY ALVEOLAR MICROLITHIASIS

Pulmonary alveolar microlithiasis (PAM) is a genetic lung disease characterized by the deposition of innumerable calcified spherules within the alveoli. These calcified spherules have been termed microliths, calcospherites, or calciferites.

The disease often manifests between the ages of 20 and 40 years, with patients presenting with slowly progressive shortness of breath and a nonproductive cough. The radiological findings are often dramatic despite the paucity of clinical symptoms. The disease is inherited in an autosomal recessive fashion, and has been linked to genetic mutations in the SLC34A2 gene that codes a type IIb sodium-phosphate cotransporter in type II alveolar cells. This abnormality leads to the accumulation of phosphate within the alveolar spaces and the development of the calcium-phosphate microliths.

PFTs findings are consistent with an intrinsic restrictive ventilatory defect. The CXR will often show diffuse involvement of the lungs, often termed "sandstorm" appearance. Chest CT will demonstrate numerous granular, sand-like calcifications throughout the lungs with subpleural and peribronchovascular distribution. Additionally, findings of crazy-paving pattern, calcified intralobular septae, and small subpleural cysts may be present. The diagnosis can confidently be made based on typical chest CT findings. In cases where lung biopsy is pursued, histological analysis will reveal concentrically laminated and calcified spheroid bodies within the alveoli (Fig. 45.3).

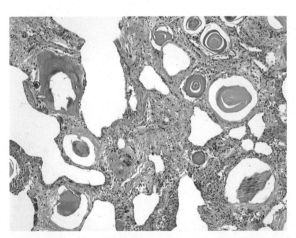

FIGURE 45.3   Histological examination demonstrating concentric lamellar calcifications in the alveolar spaces. *Courtesy of Humberto Trejo Bittar, MD.*

To date there is no effective treatment for PAM. In advanced cases, lung transplantation may be considered if respiratory failure or right heart failure develops.

TAKEAWAY POINTS

- PAM is an autosomal recessive lung disease resulting in diffuse calcified micronodules within the lungs.
- The radiological findings are often out of proportion to the patient's clinical presentation.

FURTHER READING

Ferreira Francisco, F.A., Pereira e Silva, J.L., Hochhegger, B., Zanetti, G., Marchiori, E., 2013. Pulmonary alveolar microlithiasis. State-of-the-art review. Respir. Med. 107 (1), 1–9.

Huqun, I.S., Miyazawa, H., Ishii, K., Uchiyama, B., Ishida, T., et al., 2007. Mutations in the SLC34A2 gene are associated with pulmonary alveolar microlithiasis. Am. J. Respir. Crit. Care Med. 175 (3), 263–268.

Mariotta, S., Ricci, A., Papale, M., et al., 2004. Pulmonary alveolar microlithiasis: report on 576 cases published in the literature. Sarcoidosis Vasc. Diffuse Lung Dis. 21, 173–181.

Case 46

A 29-year-old female presents to clinic after complaining of unexplained dyspnea on exertion and found to have a soft tissue density in the right lower lobe on a chest X-ray (CXR). A contrasted computed tomography (CT) scan of the chest reveals the lesion to be a vascular malformation with one feeding artery and vein (Fig. 46.1). She has no history of smoking, drug abuse, and an unrevealing family history. On further questioning, the patient denies any history of epistaxis but confirms that her dyspnea is more pronounced in the mornings after sitting up from a lying position in bed. She undergoes a bubble echocardiogram study that demonstrates delayed appearance of agitated bubbles in the left atrium after three heartbeats. Physical examination is normal without any evidence of mucocutaneous telangiectasia.

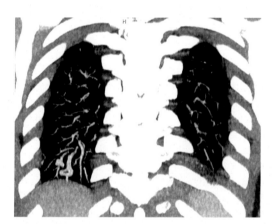

FIGURE 46.1 Contrasted coronal chest computed tomography scan showing a right lower lobe arteriovenous malformation with a single feeding pulmonary artery and drained by a single pulmonary vein.

What are the various causes of PAVM?
What is the first-line treatment of symptomatic patients?

PULMONARY ARTERIOVENOUS MALFORMATIONS

Pulmonary arteriovenous malformations (PAVM) are abnormal vascular communications between a pulmonary artery and pulmonary vein, bypassing the capillary bed, resulting in a right to left intrapulmonary shunt. Churton first described them on autopsy in 1897. PAVM frequently occur in the lower lobes, more often on the left.

The causes of PAVM are as follows:

- Hereditary hemorrhagic telangiectasia (HHT): 80–95% of all cases
- Idiopathic
- In the setting of liver disease (hepatopulmonary syndrome)
- Occurring after surgery for congenital heart disease (bidirectional cavopulmonary shunts)
- Mitral stenosis
- Penetrating chest trauma
- Infections: tuberculosis, schistosomiasis, actinomycosis
- Fanconi syndrome

Patients may be asymptomatic or present with symptoms of unexplained shortness of breath or hemoptysis depending on the degree of shunting through the PAVM. Most of the patients may experience platypnea (dyspnea on sitting up from supine position) and have evidence of orthodeoxia (hypoxemia on sitting up from supine position), due to the increased blood flow through the PAVM in the lower lobes in the upright position. On examination, there may be evidence of cyanosis and clubbing as well as systemic manifestations of underlying condition (telangiectasia, jaundice, etc.,). PAVM are associated with a risk of paradoxical embolization, and patients may present with neurological manifestations of headaches, stroke, and brain abscess.

Contrasted echocardiography is the most sensitive tool to screen for PAVM. The delayed appearance of agitated saline microbubbles in the left atrium after four cardiac cycles indicated intrapulmonary shunting. Their appearance in the left atrium within one to two cardiac cycles suggests intracardiac shunting. On CXR, large PAVM may appear as well demarcated nodule or mass. A chest CT scan may reveal a serpiginous mass or nodule with identifiable feeding arteries and draining veins. The chest CT scan may be normal in the setting of microscopic PAVM. Further testing to quantify the degree of shunting may be performed using the 100% oxygen test or a radionuclide scan using Tch-99m-labeled albumin macroaggregates.

Percutaneous transcatheter embolization is used to treat the PAVM to improve the hypoxemia and reduce the risk of paradoxical embolization. Symptomatic patients as well as those with a feeding artery diameter ≥2 mm on chest CT should be referred to determine feasibility of embolization. In cases of severe hemoptysis symptoms, persistent symptoms despite repeat embolization, or where the embolization can be performed; surgical lung resection or lung transplantation may be required. PAVM may reoccur after embolization or surgery and patients should be followed up treatment accordingly. Antibiotic prophylaxis is recommended for procedures with risk of bacteremia.

HEREDITARY HEMORRHAGIC TELANGIECTASIA

HHT, also known as Osler-Weber-Rendu syndrome, is an autosomal dominant vascular disorder characterized by mucocutaneous telangiectasia, recurrent epistaxis, and visceral arteriovenous malformations. It is the most common cause of PAVM and has been linked to mutations in the ENG, ACVRL1, and SMAD4 genes. Diagnosis is made or suspected based on the Curaçao criteria. Genetic screening and counseling should be offered to patients with HHT.

TAKEAWAY POINTS

- HHT is the most common cause of PAVM.
- The majority of PAVM occur in the lower lobes, giving rise to unique clinical symptoms and signs such as platypnea and orthodeoxia.
- Contrasted echocardiography is the most sensitive test to detect PAVM.

FURTHER READING

Cartin-Ceba, R., Swanson, K.L., Krowka, M.J., 2013. Pulmonary arteriovenous malformations. Chest 144 (3), 1033–1044.

Churton, T., 1897. Multiple aneurysm of pulmonary artery. BMJ 1, 1223.

Dines, D.E., Arms, R.A., Bernatz, P.E., Gomes, M.R., 1974. Pulmonary arteriovenous fistulas. Mayo Clin. Proc. 49 (7), 460–465.

van Gent, M.W., Post, M.C., Luermans, J.G., Snijder, R.J., Westermann, C.J., Plokker, H.W., et al., 2009. Screening for pulmonary arteriovenous malformations using transthoracic contrast echocardiography: a prospective study. Eur. Respi. J. 33 (1), 85–91.

McDonald, J., Bayrak-Toydemir, P., Pyeritz, R.E., 2011. Hereditary hemorrhagic telangiectasia: an overview of diagnosis, management, and pathogenesis. Genet. Med. 13 (7), 607–616.

Case 47

A 68-year-old man presents with acute and progressive shortness of breath and a 15-pound weight loss over the last 2 months, along with fatigue and leg swelling. He has a 30 pack-year smoking history and works as an architect. On examination, he is hypoxic and is noted to have jugular venous distention, a loud P2 on cardiac auscultation and bilateral lower leg pitting edema. Lung auscultation is unremarkable.

He undergoes a contrasted chest computed tomography (CT) scan that shows bilateral diffuse centrilobular ground-glass nodules, an enlarged pulmonary artery and right cardiac chambers, and no evidence of pulmonary embolism. An echocardiogram reveals an elevated right ventricle systolic pressure and dilated right atrium and right ventricle. A right heart catheterization reveals hemodynamics consistent with precapillary pulmonary hypertension. A ventilation-perfusion scan of the lungs showed bilateral, peripheral subsegmental defects with ventilation perfusion mismatch.

The patient's dyspnea and hypoxia continued to rapidly progress during his hospital stay and he eventually died a week after his admission. Autopsy revealed a poorly differentiated gastric adenocarcinoma. Tumor cells were found within the small pulmonary arteries and arterioles that were also consistent with gastric adenocarcinoma and stained positive for vascular endothelial growth factor (VEGF) and tissue factor. Also noted were microthrombi and intimal fibrocellular proliferation involving the pulmonary arteries and arterioles.

What is the diagnosis?

PULMONARY TUMOR THROMBOTIC MICROANGIOPATHY

Pulmonary tumor thrombotic microangiopathy (PTTM), first described by von Herbay in 1990, is a pathophysiological entity characterized by the presence of tumor cell emboli in the subsegmental pulmonary arteries and arterioles with associated microthrombus formation and diffuse intimal myofibroblast proliferation resulting in pulmonary vessel occlusion.

Occult gastric adenocarcinoma is the most associated tumor with this syndrome. Patients present with acute progressive dyspnea and cough and later progress to pulmonary hypertension, right heart failure, and death. The chest CT findings are nonspecific and include findings of centrilobular ground-glass opacities, septal thickening, mediastinal adenopathy, and tree-in-bud opacities. Findings consistent with pulmonary hypertension and right heart strain may be seen as evidenced by an enlarged pulmonary artery and right cardiac chamber dilatation. The hemodynamics on right heart catheterization are often consistent with precapillary pulmonary hypertension. Ventilation-perfusion lung scanning will reveal multiple peripheral subsegmental defects, with V/Q mismatch, giving a checkerboard appearance, termed the checkerboard sign.

Patients with PTTM have a very poor prognosis and the diagnosis is often made on autopsy. Antemortem diagnosis can be made on finding intravascular tumor cells with associated thrombi and intimal fibrocellular proliferation on lung biopsy (transbronchial or video-assisted thoracoscopic surgery), or on finding tumor cells on wedged pulmonary artery catheterization aspiration. Positive tumor cell immunohistochemical staining for vascular endothelial growth factor and tissue factor along with the intimal proliferation aids in confirming the diagnosis of PTTM and distinguishing it from pulmonary tumor emboli syndrome. Pulmonary tumor emboli syndrome, a similar condition in which tumor cells occlude the pulmonary vessels, is distinguished by the absence of the associated intimal proliferation seen in PTTM.

There is no definite treatment, although a few reports suggest that systemic chemotherapy including imatinib may prolong survival. Overall, the prognosis is grave with average interval between dyspnea onset and death being less than 2 weeks.

TAKEAWAY POINTS

- PTTM should be considered in patients with acute unexplained dyspnea and rapidly progressive pulmonary hypertension.
- Positive cytology staining for tumor cells from the wedged pulmonary catheter aspirate may assist in making the diagnosis antemortem; however, prognosis remains poor.

FURTHER READING

von Herbay, A., Illes, A., Waldherr, R., Otto, H.F., 1990. Pulmonary tumor thrombotic microangiopathy with pulmonary hypertension. Cancer 66 (3), 587–592.

Higo, K., Kubota, K., Takeda, A., Higashi, M., Ohishi, M., 2014. Successful antemortem diagnosis and treatment of pulmonary tumor thrombotic microangiopathy. Intern. Med. 53 (22), 2595–2599.

Uruga, H., Fujii, T., Kurosaki, A., Hanada, S., Takaya, H., Miyamoto, A., et al., 2013. Pulmonary tumor thrombotic microangiopathy: a clinical analysis of 30 autopsy cases. Intern. Med. 52 (12), 1317–1323.

Case 48

A 42-year-old male presents with a nonproductive cough and choking sensation in the neck that started a few months prior. He is smoker with a history of recurrent pneumonias in the right lower lobe. A chest X-ray (CXR) is done showing an infiltrate in the right lower lobe; a chest computed tomography (CT) scan confirms the presence of the infiltrate with evidence of volume loss. Bronchoscopic examination shows multiple exophytic papillary growth on the vocal cords and the superior segment of the right lower lobe (Fig. 48.1). Biopsy of the lesion revealed fragments of polypoid tissue with dense fibrovascular cores (Fig. 48.2).

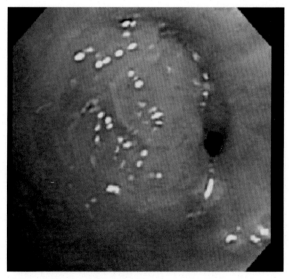

FIGURE 48.1 Bronchoscopic examination of the airway revealing a polypoid intraluminal lesion in the superior segment of the right lower lobe.

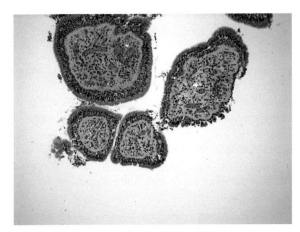

FIGURE 48.2 Histology revealing polypoid tissue with dense fibrovascular core.

What is the diagnosis?
What treatment options do the patient have?

RECURRENT RESPIRATORY PAPILLOMATOSIS

Recurrent respiratory papillomatosis (RRP) is a rare disorder characterized by recurrent benign exophytic wart-like growths (papillomas) involving the epithelium mucosa of the upper and lower respiratory tract secondary to human papillomavirus (HPV) infection.

Two forms of the disease exist: the juvenile-onset RRP and adult-onset RRP. Juvenile-onset RRP is usually diagnosed in children less than 5 years of age and is presumed to be transmitted from an infected mother to the infant during childbirth. Adult-RRP transmission might be related to oral sex, and the most significant risk factor is the total number of lifetime sexual partners of the patient. RRP has a predilection to involve the upper respiratory tract and larynx in 95% of cases. In 5% of cases the trachea may be affected, while involvement of the lung parenchymal occurs in <1% of cases. HPV-6 and HPV-11 account for over 90% of RRP cases, with HPV-11 resulting in more aggressive disease and a higher risk of malignant transformation.

Patients often present with hoarseness, foreign body sensation in the throat, dysphagia, coughing, and dyspnea. In cases of airway obstruction, inspiratory wheezing, stridor, and postobstructive pneumonia may occur. The CXR is often unrevealing. The chest CT scan may reveal multiple intratracheal growths, parenchymal nodules, and cysts. Diagnosis is often made on bronchoscopic evaluation revealing findings of cauliflower-like growths in the upper and lower respiratory tract. Histological examination of the lesions will show projections of nonkeratinized stratified squamous epithelium covering a fibrovascular core. HPV DNA may be detected by polymerase chain reaction and the specific type determined if needed.

Malignant transformation to squamous cell carcinoma may occur in 3–5% of cases with parenchymal involvement. There is no cure for RRP since the lesions almost always recur after resection or treatment. Patients are managed with repeated debulking of the airways lesions (carbon dioxide or argon laser, cryotherapy, or surgical excision). Respiratory precautions should always be taken by the practitioners during bronchoscopy to prevent exposure and transmission. Injection of cidofovir into the base of a lesion after excision may slow regrowth. Other medical treatments described include the use of interferon α-2a, bevacizumab, ribavirin, methotrexate, and acyclovir. Prophylactic HPV vaccine has been shown to decrease the interval between surgical procedures; however, further studies to show its efficacy in RRP are required.

TAKEAWAY POINTS

- RRP has a bimodal age distribution with HPV-6 and HPV-11 causing the majority of cases.
- The larynx is most commonly affected. It is unusual for RRP to manifest in the lower respiratory tracts without laryngeal involvement.

FURTHER READING

Goon, P., Sonnex, C., Jani, P., Stanley, M., Sudhoff, H., 2008. Recurrent respiratory papillomatosis: an overview of current thinking and treatment. Eur. Arch. Otorhinolaryngol. 265 (2), 147–151.

Shehab, N., Sweet, B.V., Hogikyan, N.D., 2005. Cidofovir for the treatment of recurrent respiratory papillomatosis: a review of the literature. Pharmacotherapy 25 (7), 977–989.

Case 49

A 62-year-old male, with history of congestive heart disease, presents with short-ness of breath and symptoms consistent with an acute heart failure exacerba-tion. Chest imaging reveals a right-sided pleural effusion and calcifications in the right lower lobe. His chest computed tomography scan is shown in Fig. 49.1. The patient is medically managed, and once stable, undergoes bronchoscopy with transbronchial biopsies of the right lower lobe. The lung histology reveals the presence of mature bone fragments within the alveoli (Fig. 49.2).

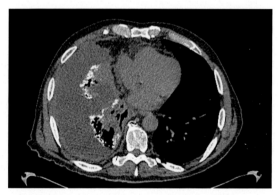

FIGURE 49.1 Chest computed tomography scan showing subpleural micronodular lesions mainly in the right lower lobe.

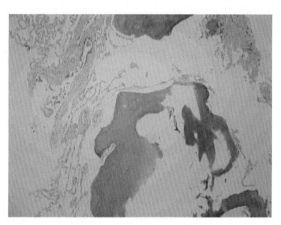

FIGURE 49.2 Lung biopsy showing bone tissue located within the alveolar spaces with bone mar-row identified within some of the fragments. No evidence of granulomatosis or malignancy is present.

What is the diagnosis?
What are the different types and etiologies of this condition?

DIFFUSE PULMONARY OSSIFICATION

Pulmonary ossification is characterized by metaplastic bone formation, often containing marrow elements, within the lung parenchyma. It was first described by Luschka in 1856. Pulmonary ossification may be either (1) diffuse, or more often (2) localized, resulting in dystrophic calcification within lung scar tissue, granulomas, or bronchial cartilage.

There are two types of diffuse of diffuse pulmonary ossification (DPO):

- Dendriform: branching calcified bone deposits along the terminal airways and alveolar septa. Although this may be idiopathic, it is more commonly associated with underlying lung disease (pulmonary fibrosis, recurrent pneumonias, pneumoconiosis, asbestosis, acute respiratory distress syndrome).
- Nodular: the more common type that is characterized by small (<1 cm) circumscribed lamellar bone deposits within the alveoli spaces that do not contain marrow elements. It has been associated with congestive heart disease, mitral stenosis, and hypertrophic subvalvular aortic stenosis.

Hypercalcemia, hyperphosphatemia, alkalosis, and underlying lung injury are all factors that support pulmonary ossification formation. Parenchymal scar tissue secondary to lung inflammation and injury creates an alkaline environment that favors calcium deposition, alkaline phosphatase activation, and the triggering of profibrogenetic cytokines that enhance osteoblastic activity and bone formation.

The diagnosis of DPO is often made incidentally or on postmortem examination. It is seen more frequently in men in their fifth to sixth decade, who are often asymptomatic or complain of nonspecific symptoms such as a chronic cough. Pulmonary function tests are usually normal but a restrictive physiology with low diffusion capacity may be seen in extensive cases. Serum calcium, phosphate, and alkaline phosphatase levels are normal.

Patients may be misdiagnosed with pulmonary fibrosis, bronchiectasis, or lymphangitic tumor spread on imaging. High-resolution computed tomography will show asymmetrical subpleural calcified reticulonodular densities, most commonly in the lung bases. Dendriform lesions will have a branching pattern in a bronchovascular distribution, while nodular lesions appear as multiple calcified nodules that may be mistaken for calcified granulomas. Ossification may also be detected on [99m]technetium-methylene diphosphate bone scintigraphy.

Lung biopsy (open biopsy or transbronchial) will reveal the presence of mature bone spicules within the lung parenchyma, often in association with lung fibrosis. Bone marrow may be present within these fragments.

No treatment guidelines exist. Management is supportive and directed at treating the underlying cardiac or pulmonary disease. No benefit has been seen with use of low calcium diets and systemic steroid for treating DPO. Reports of a potential role for bisphosphonates and warfarin in preventing progression still need to be proven.

TAKEAWAY POINTS

- DPO may be idiopathic but more commonly occurs in the setting of underlying cardiac, pulmonary, or systemic disorders.
- DPO is often an incidental finding with symptoms, when present, related to the underlying disorder.

FURTHER READING

Chan, E.D., Morales, D.V., Welsh, C.H., McDermott, M.T., Schwarz, M.I., 2002. Calcium deposition with or without bone formation in the lung. Am. J. Respir. Crit. Care Med. 165 (12), 1654–1669.

Lara, J.F., Catroppo, J.F., Kim, D.U., da Costa, D., 2005. Dendriform pulmonary ossification, a form of diffuse pulmonary ossification: report of a 26-year autopsy experience. Arch. Pathol. Lab. Med. 129 (3), 348–353.

Luschka, H., 1856. Ramified ossification of the pulmonary parenchyma. Wirchows Arch. 10, 500–505.

Case 50

A 28-year-old female presents complaining of a chronic productive cough. Her medical history is relevant for recurrent otitis media, sinusitis, and infertility. She has no relevant family history and is a nonsmoker. She has been bringing up thick dark sputum for the last 6 months. On examination, she has bibasilar coarse crackles on lung auscultation, distant heart sounds, and digital clubbing. A chest X-ray (CXR) and chest computed tomography (CT) are performed (Figs. 50.1 and 50.2). Testing for cystic fibrosis comes back negative and her immunoglobulin levels are within normal range.

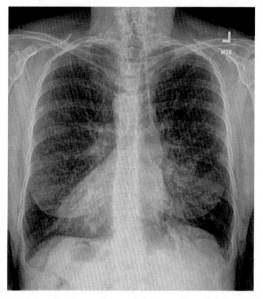

FIGURE 50.1   Chest X-ray showing evidence of dextrocardia and right gastric bubble. There is increased interstitial markings.

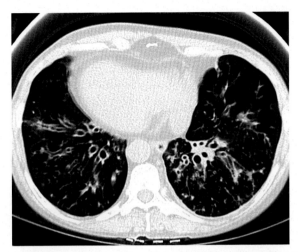

FIGURE 50.2 Chest computed tomography scan showing dextrocardia and bronchiectasis of the lower lobes.

What is the diagnosis?

PRIMARY CILIARY DYSKINESIA

Primary ciliary dyskinesia (PCD), or immotile-cilia syndrome, is a heterogeneous disorder in which impaired motility of airway cilia results in impaired mucociliary clearance. It is a genetic disorder with autosomal recessive inheritance. The disease is characterized by chronic sinopulmonary infections, chronic otitis media, bronchiectasis, male infertility, and situs inversus in 50% of cases. The triad of situs inversus, bronchiectasis, and sinusitis was first described in 1933 and is termed Kartagener syndrome. It was not till over 40 years later that the link to defective cilia structure and function was determined.

PCD manifestations occur as a result of partial or total absence of the outer and/or inner dynein arms of the cilia that line the respiratory tract and internal organs. Numerous genes have been implicated in PCD; however, DNAI1 and DNAH5 genes have been found to be involved in up to 38% of cases.

The disease often manifests in infancy as respiratory insufficiency or tachypnea but may also first present in adulthood. Affected individuals will have complaints of chronic cough, recurrent otitis media, chronic sinusitis, recurrent pneumonias, and bronchiectasis. Patients may also complain of hearing loss and a poor sense of smell. Less than 10% may have pectus excavatum and scoliosis. Affected males will be infertile while females will have difficulty conceiving. Due to abnormal ciliary motility during development, half of the patients with PCD may have situs inversus. Approximately 6% of individuals with PCD may have situs ambiguus, where the organs are neither in the typical position (situs solitus) nor entirely reversed (situs inversus totalis). This subset of patients is more likely to have underlying congenital heart disease than the general population.

Pulmonary function testing may reveal a mild to moderate obstructive ventilatory defect. CXR may be normal or show dextrocardia and bronchiectasis. Chest CT imaging findings may include peribronchial thickening, peripheral mucus plugging, bronchiectasis, and situs inversus. Multiple tests have been developed in order to screen and diagnose patients with PCD. Measuring the levels of nasal nitric oxide (NO) have found to be a good screening tool in which patients with PCD will be commonly found to have low nasal NO levels. A nasal epithelium biopsy and examination with transmission electron microscopy and high-speed video microscopy can examine the ciliary ultrastructure and function and confirm the diagnosis of PCD. It should be noted, however, that 30% of patients with PCD may have normal ciliary structure on examination; as such, the use of various different tests to reach the diagnosis may be needed. Finally, genetic testing to determine the presence of specific mutations can be performed at specialized centers and is diagnostic when biallelic mutations are present.

Management of PCD is directed at managing impaired mucus clearing with chest physiotherapy and airway clearance techniques, as well as treating the chronic and recurrent infections. Surgical intervention may be required in cases of chronic otitis media and sinusitis. Lung transplantation may be considered for advanced cases of bronchiectasis.

TAKEAWAY POINTS

• Consider PCD in patients with chronic sinopulmonary symptoms and bronchiectasis once cystic fibrosis is ruled out.
• While there is no gold standard test for PCD, diagnosis may be made by nasal NO measurement, nasal biopsy with electron microscopy, or genetic testing.

FURTHER READING

Kartagener, M., 1933. Zur pathogenese der bronkiectasien. Bronkiectasien bei situs viscerum inversus. Beitr Klin Tuberk Spezif Tuberkuloseforsch 83, 489–501.

Knowles, M.R., Daniels, L.A., Davis, S.D., Zariwala, M.A., Leigh, M.W., 2013. Primary ciliary dyskinesia. Recent advances in diagnostics, genetics, and characterization of clinical disease. Am. J. Respir. Crit. Care Med. 188 (8), 913–922.

Zariwala, M.A., Knowles, M.R., Omran, H., 2007. Genetic defects in ciliary structure and function. Annu. Rev. Physiol. 69, 423–450.

Case 51

A 65-year-old male with history of multiple myeloma (IgG subtype), diagnosed 1 year ago, and obstructive sleep apnea presents with fatigue, a nonproductive cough, and shortness of breath on exertion. His myeloma was initially treated with corticosteroids and melphalan and subsequently with bortezomib. His vital signs and physical examination are unremarkable. A chest X-ray shows a small to moderate left-sided pleural effusion. A chest computed tomography scan confirms the presence of the pleural effusion without any underlying lung parenchymal or bone abnormalities (Fig. 51.1). Echocardiogram is normal with a left ventricular ejection fraction of 55%.

The patient undergoes a thoracentesis that demonstrates a dark red, lymphocytic predominant (70%) exudative pleural effusion with a pleural glucose of 90 mg/dL and pleural pH of 7.50. Pleural cytology with flow cytometry shows monoclonal plasma cells that are positive for CD138, and negative for calretinin, CK AE1/AE3, CK7, CK20, and TTF-1.

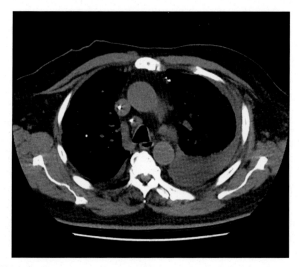

FIGURE 51.1   Chest computed tomography scan showing left-sided pleural effusion.

What is the diagnosis?

MYELOMATOUS PLEURAL EFFUSION

Pleural effusions occur in 6% of multiple myeloma cases and are normally benign and secondary to congestive heart failure, chronic renal disease, amyloidosis, and infections. Multiple myeloma involvement of the pleural space occurs in less than 1% of cases and is a late complication of myeloma and a sign of aggressive disease. The patient's history of multiple myeloma, presentation, and pleural fluid cytology analysis in this case is consistent with myelomatous pleural effusion (MPE).

The exact mechanism of MPE pathogenesis remains unclear, but proposed theories include myeloma cell infiltration or invasion from adjacent skeletal lesions, extension from chest wall plasmacytomas, tumor infiltration of the pleura, or mediastinal lymph node involvement resulting in lymphatic obstruction.

Patients may be asymptomatic or present with symptoms of chest pain, fatigue, nonproductive cough, shortness of breath, and fevers. They often have marked increase in serum β-2-microglobulin, lactate dehydrogenase, and C-reactive protein at the time of pleural effusion presentation in comparison to their initial presentation with multiple myeloma. In a few cases, MPE may be the first presentation of multiple myeloma. The most common multiple myeloma type to be associated with MPE is the IgA multiple myeloma subtype.

The pleural effusions are exudates and can be unilateral, with left-sided effusions reported more frequently than right, or bilateral. MPE may be associated with pleural or chest wall plasmacytomas or pulmonary parenchymal lesions. Diagnosis is established in multiple myeloma patients based on pleural fluid cytology and flow cytometry. The pleural fluid cytology is positive for myelomatous involvement in the majority of cases. Pleural fluid adenosine deaminase levels may be elevated in some cases without any evidence of pulmonary tuberculosis. In cases where pleural fluid cytology is negative, thoracoscopic evaluation with pleural biopsy may confirm the diagnosis.

Treatment involves systemic chemotherapy for the multiple myeloma and pleurodesis of the pleural space. Despite treatment and pleurodesis, approximately one-third of effusions recur within a few months. The median reported survival is 4 months from the onset of MPE.

TAKEAWAY POINTS

- MPE is a rare complication of multiple myeloma, diagnosed on pleural cytology and flow cytometry.
- MPE effusions are exudates and commonly left sided. They indicate an aggressive disease course and poor survival.

FURTHER READING

Kamble, R., Wilson, C.S., Fassas, A., Desikan, R., Siegel, D.S., Tricot, G., et al., 2005. Malignant pleural effusion of multiple myeloma: prognostic factors and outcome. Leuk. Lymphoma 46 (8), 1137–1142.

Kintzer Jr., J.S., Rosenow 3rd, E.C., Kyle, R.A., 1978. Thoracic and pulmonary abnormalities in multiple myeloma. A review of 958 cases. Arch. Intern. Med. 138 (5), 727–730.

Case 52

A 34-year-old female is seen with symptoms of nonproductive cough and wheezing for the last year. She was diagnosed with asthma a year ago, but has not experienced any improvement in her symptoms with her prescribed inhalers. She has a history of recurrent outer ear cartilage and ocular inflammation. On examination, the patient has cauliflower appearing ears, saddle-shaped nose, thyroid cartilage tenderness on palpation, and is noted to have inspiratory and expiratory wheezing on chest auscultation.

Pulmonary function tests (PFTs) reveal a mild obstructive ventilatory defect with no response to bronchodilators. Inspiratory and expiratory flow-volume loops show evidence of variable intrathoracic upper airway obstruction. A dynamic chest computed tomography (CT) is performed that reveals tracheal collapse on expiration, associated with tracheal wall thickening and calcification that spares the posterior membranous portion of the trachea.

What is the patient's diagnosis?
What is MAGIC syndrome?

RELAPSING POLYCHONDRITIS

Relapsing polychondritis (RP) is an autoimmune condition of unclear etiology in which patients develop antibodies directed mainly against type II collagen. The disease is characterized by multisystem episodes of relapsing cartilage inflammation, predominately affecting the ears, nose, larynx, and tracheobronchial tree. It was first described by Jaksch-Wartenhorst in 1923, and initially termed polychondropathia.

Several diagnostic criteria have been proposed, with the most recent modification for the clinical diagnosis of RP, including the following:

The presence of two major criteria, or of one major and two minor criteria:

- Major criteria:
  - Auricular cartilage inflammation
  - Inflammation of the nasal cartilage
  - Inflammation of the laryngotracheobronchial tree
- Minor criteria:
  - Seronegative arthritis
  - Ocular inflammation
  - Hearing change/loss
  - Vestibular dysfunction

Inflammation of the ear cartilage is the most common presentation, with patients experiencing relapsing episodes of external ear inflammation that over time may lead to the development of a cauliflower ear. The nasal cartilage is the second most involved site, with recurrent inflammatory episodes resulting in the destruction of the nasal septum and the development of saddle-nose deformity. Involvement of the respiratory tract cartilage may occur in 20–50% of patients during the course of their disease. This is seen more commonly in women, and may be severe and life threatening. Laryngeal chondritis may lead to hoarseness, voice loss, and in severe cases laryngeal stenosis. Patients with tracheobronchial chondritis usually complain of progressive dyspnea, nonproductive cough, recurrent infections, and wheezing. They may have tenderness to palpation of the anterior trachea and thyroid cartilage. Development of tracheobronchomalacia with expiratory airway collapse, may in severe cases, progress to tracheal or bronchial stenosis.

One-third of cases of RP occur in association with other autoimmune conditions, vasculitis or hematological disorders. The overlap of Behçet disease and RP has been termed MAGIC syndrome (mouth and genital ulcers with inflammatory cartilage).

Patients may have elevated serum inflammatory markers. PFTs may show an obstructive ventilatory defect. Intrathoracic and/or extrathoracic upper airway obstruction may be evident on analysis of the flow-volume loop. Chest CT imaging may show airway wall thickening, airway narrowing, and calcification of the airway cartilage that spares the posterior membranous wall. Dynamic chest CT imaging may demonstrate evidence of expiratory airway collapse in patients with tracheobronchomalacia. Tissue biopsy is not required to make the diagnosis, but if done will demonstrate underlying chondritis, perichondritis, or chondrolysis.

There are no official guidelines for the management of RP, although there are many case reports and small series of patients reporting various treatment approaches. Patients are usually managed with nonsteroidal anti-inflammatory agents or corticosteroids for acute episodes and may require steroid-sparing immunosuppressive agents in severe cases. Bronchial stenosis may be managed with bronchoscopic interventions including airway dilation or stenting. Patients with laryngeal stenosis may require tracheostomy.

TAKEAWAY POINTS

- The respiratory tract may be involved in 50% of patients with RP and is a significant cause of morbidity and mortality.
- RP is a multisystem disease that often occurs in association with other systemic conditions and may have a variety of clinical presentations depending on the site of the involved cartilage.

FURTHER READING

Firestein, G.S., Gruber, H.E., Weisman, M.H., Zvaifler, N.J., Barber, J., O'Duffy, J.D., 1985. Mouth and genital ulcers with inflamed cartilage: MAGIC syndrome. Five patients with features of relapsing polychondritis and Behcet's disease. Am. J. Med. 79 (1), 65–72.

Michet Jr., C.J., McKenna, C.H., Luthra, H.S., O'Fallon, W.M., 1986. Relapsing polychondritis. Survival and predictive role of early disease manifestations. Ann. Intern. Med. 104 (1), 74–78.

Jaksch-Wartenhorst, R., 1923. Polychondropathia. Wien Arch. F Inn. Med. 6, 93–100.

Rafeq, S., Trentham, D., Ernst, A., 2010. Pulmonary manifestations of relapsing polychondritis. Clin. Chest Med. 31 (3), 513–518.

Case 53

A 25-year-old male presents with complaints of swelling on the lateral aspect of his neck and a mild nonproductive cough. He has not noticed any swelling elsewhere and does not endorse any B-symptoms. Examination reveals multiple painless and enlarged cervical lymph nodes. His chest X-ray shows evidence of mediastinal adenopathy but is otherwise normal.

A biopsy of the cervical lymph nodes reveals lymph node sinus enlargement and infiltration by histiocytes that stain positive for S100 and CD68 but negative for CD1a. There is also evidence of emperipolesis on histology.

What is the diagnosis?

ROSAI-DORFMAN DISEASE

Rosai-Dorfman disease (RDD), also known as sinus histiocytosis with massive lymphadenopathy, is a benign histiocyte proliferative disorder of unclear etiology that usually manifests as painless lymphadenopathy. Rosai and Dorfman first described the disorder in 1969.

While the exact etiology of the disease remains unknown, it is hypothesized that a cytokine-mediated process of histiocyte accumulation within the lymph node occurs as a result of an underlying autoimmune disorder, infectious process (Epstein-Barr virus and Human Herpes virus-6), or malignancy. Dysfunctional apoptosis as a result of defective Fas/FasL signaling may also be involved in histiocyte proliferation.

There is histiocytic expansion within the lymph node sinuses of patients with RDD. Commonly involved lymph nodes are neck, axilla, inguinal area, and mediastinum. Extranodal involvement occurs as a result of lymphatic involvement with RDD, and occurs in 20–40% of cases. Extranodal RDD has been reported to involve skin, sinus and oral cavities, respiratory tract, gastrointestinal system, and central nervous system.

Intrathoracic involvement may manifest as:

- Mediastinal and hilar adenopathy
- Interstitial lung disease
- Cystic lung disease
- Pleural effusions
- Airway involvement
- Pulmonary nodules
- Pulmonary artery involvement

The disease often presents in the first two decades of life as painless cervical adenopathy. Patients may present with respiratory symptoms depending on the site of intrathoracic involvement. Chest computed tomography findings vary and may include cervical and thoracic adenopathy that are hypermetabolic on positron emission tomography scan imaging. Histologically, there is lymph node pericapsular fibrosis and sinus infiltration and expansion by lymphoplasmacytic cells and histiocytes. Emperipolesis (histiocyte engulfment of lymphocytes and erythrocytes) is evident and is considered as a hallmark of the disease. Immunohistochemical staining is positive for S100, CD68, CD163, alpha-1 antichymotrypsin, and negative for CD1a.

RDD commonly follows an indolent course, so that patients with localized adenopathy are followed clinically. Approximately 20% of patients may have spontaneous remission of the disease. In patients with organ involvement and significant symptoms, treatment is directed at symptom management and may include surgical debulking of lymph nodes causing compression of adjacent structures. Systemic corticosteroids, immunosuppressive agents, and radiotherapy have been tried for severe disease; however, their efficacy remains unclear.

TAKEAWAY POINTS

- RDD is a histiocytic disorder that commonly presents as painless cervical lymphadenopathy. Mediastinal adenopathy is the most common thoracic manifestation.
- Diagnosis is made on tissue biopsy that reveals emperipolesis and histiocytes that stain positive for S100 and negative for CD1a.

FURTHER READING

Cartin-Ceba, R., Golbin, J.M., Yi, E.S., Prakash, U.B., Vassallo, R., 2010. Intrathoracic manifestations of Rosai–Dorfman disease. Respir. Med. 104 (9), 1344–1349.

Pulsoni, A., Anghel, G., Falcucci, P., Matera, R., Pescarmona, E., Ribersani, M., et al., 2002. Treatment of sinus histiocytosis with massive lymphadenopathy (Rosai–Dorfman disease): report of a case and literature review. Am. J. Hematol. 69 (1), 67–71.

Rosai, J., Dorfman, R.F., 1969. Sinus histiocytosis with massive lymphadenopathy. A newly recognized benign clinicopathological entity. Arch. Pathol. 87 (1), 63–70.

Case 54

A 27-year-old female presents with acute onset shortness of breath that started 2 days ago, along with a low-grade fever and a nonproductive cough. She has no significant medical history and does not smoke or use recreational drugs. She is an office worker and denies recent travel. She states that she was at a silicone pumping party the day before the onset of her symptoms. On examination the patient is tachycardic, tachypneic, and hypoxic on room air. She has bilateral expiratory wheezes on auscultation of the chest.

A chest X-ray and chest computed tomography scan are performed showing evidence of peripheral, patchy bilateral lung opacities. She undergoes bronchoscopic evaluation that is unremarkable, lavage cultures are all negative and a workup for vasculitis and other autoimmune conditions are normal. She undergoes a video-assisted thoracoscopic surgery wedge biopsy of the right lung. The histology reveals silicone vacuoles in the alveolar spaces and interalveolar walls.

What is the diagnosis?

Rare and Interesting Cases in Pulmonary Medicine.

SILICONE EMBOLISM SYNDROME

The use of liquid silicone (polydimethylsiloxane), an inert synthetic polymer, is becoming more widely used for breast augmentation and cosmetic procedures. Adverse pulmonary consequences are often the result of illicit subcutaneous silicone injections that are inadvertent injected directly into a vein resulting in acute pneumonitis, diffuse alveolar hemorrhage, and acute respiratory distress syndrome.

Presentation is very similar to fat embolism syndrome, with fever, dyspnea, cough, hypoxia, respiratory distress, hemoptysis, and altered mental status occurring within the first 72 h after an injection. Imaging may reveal bilateral peripheral patchy ground-glass infiltrates. Pulmonary function tests may be normal or have a restrictive pattern and bronchoalveolar lavage analysis will show a neutrophil predominant lavage. Lung biopsy shows silicone vacuoles in the alveolar spaces, interalveolar walls, pulmonary capillaries, and within alveolar macrophages.

Treatment is supportive. Steroids have been used in some cases but their benefit remains uncertain. Involvement of the central nervous system is associated with a higher mortality.

CHRONIC SILICONE EMBOLISM SYNDROME

Another form of pulmonary disease secondary to silicone embolism is chronic silicone embolism syndrome. It is associated with silicone breast implants, manifesting years after the implantation surgery. Asymmetry of the breast implants may be noted on examination or imaging. As in acute silicone embolization syndrome, lung histology will show silicone microemboli within the alveolar capillaries, interstitium and alveolar spaces, in association with histiocytes and multinucleated giant cells. Removal of the implants is recommended to prevent progression.

TAKEAWAY POINTS

- Silicone embolism syndrome should be suspected in any person with recent cosmetic procedures involving silicone injection and a presentation similar to fat embolism syndrome.
- The chronic form of the disease rarely occurs in patients with silicone breast implants.

FURTHER READING

Gopinath, P.P., Ali, A., Van Tornout, F., Kamath, A., Crawford, M., Nicholson, A.G., 2015. Chronic silicone embolism syndrome due to PIP breast implant leakage - a new entity? Histopathology 66 (6), 904–906.

Restrepo, C.S., Artunduaga, M., Carrillo, J.A., Rivera, A.L., Ojeda, P., Martinez-Jimenez, S., et al., 2009. Silicone pulmonary embolism: report of 10 cases and review of the literature. J. Comput. Assist. Tomogr. 33 (2), 233–237.

Zamora, A.C., Collard, H.R., Barrera, L., Mendoza, F., Webb, W.R., Carrillo, G., 2009. Silicone injection causing acute pneumonitis: a case series. Lung 187 (4), 241–244.

Case 55

A 52-year-old male presents with a nonproductive cough of 3 months duration. During this period, he has been prescribed multiple courses of antibiotics for a recurring left leg infection. He also reports chronic diarrhea. A chest X-ray and chest computed tomography scan confirm the presence of anterior mediastinal mass that on biopsy is found to be a thymoma. Given the patients relapsing left leg infection and diarrhea, he undergoes a workup for infection that does not reveal any pathogens. Further testing reveals panhypogammaglobulinemia and low CD4 T-cell counts.

What is the diagnosis?
What are other conditions associated with thymomas?

GOOD SYNDROME

Good syndrome is a paraneoplastic syndrome characterized by combined B-cell and T-cell immunodeficiency in patients with thymomas, rendering them susceptible to encapsulated bacterial infections and opportunistic infections. Dr. Robert Good first reported the association between thymoma and hypogammaglobulinemia in 1954.

The etiology of the immunodeficiency in Good syndrome remains unclear but seems to arise in the bone marrow where B-cell development is arrested. It is estimated that 6–11% of patients with thymomas have hypogammaglobulinemia. Affected adults often present between 40 and 70 years of age. These patients exhibit defects in their B-cells and T-cells, resulting in low circulating levels of B-cells, low serum immunoglobulin levels, CD4 lymphocytopenia, cell-mediated immunity defects, and recurrent upper and lower respiratory tract infections and a predisposition for opportunistic infections. Patients frequently have diarrhea that may be as a result of mucosal lesions resembling villous atrophy resulting in malabsorption or perhaps making these patients susceptible to gastrointestinal pathogens. Symptoms associated with mass affect of the thymoma may also be present. Up to 30–40% of patients are diagnosed simultaneously with thymoma and hypogammaglobulinemia.

The immunodeficiency associated with thymomas is usually not responsive to corticosteroids and other immunosuppressive therapies. Patients may require thymectomy in order to prevent progression of the underlying disease and intravenous immunoglobulin (IVIG) therapy to restore low immunoglobulin levels. Patients with persistent infections may require long-term antibiotic therapy in some cases.

OTHER PARANEOPLASTIC/AUTOIMMUNE CONDITIONS ASSOCIATED WITH THYMOMAS

- Myasthenia gravis
- Lambert-Eaton syndrome
- Pancytopenia
- Pure red-cell aplasia
- Autoimmune disorders (systemic lupus erythematosus, mixed connective tissue disease, thyroiditis)
- Autoimmune enteropathy
- Limbic encephalitis
- Neuromyotonia (Isaac syndrome)
- Stiff person syndrome
- Myotonic dystrophy
- Lichen planus
- Pemphigus vulgaris

TAKEAWAY POINTS

- Good syndrome describes the adult-onset immunodeficiency with low immunoglobulin levels in the setting of thymomas. The conditions often present simultaneously.
- Treatment is done with thymectomy and IVIG therapy.

FURTHER READING

Good, R.A., 1954. Agammaglobulinemia—a Provocative Experiment of Nature. 26. Bull University Minn Hosp Minn Med Fdn, pp. 1–19.

Kelesidis, T., Yang, O., 2010. Good's syndrome remains a mystery after 55 years: a systematic review of the scientific evidence. Clin. Immunol. (Orlando, Fla) 135 (3), 347–363.

Tarr, P.E., Sneller, M.C., Mechanic, L.J., Economides, A., Eger, C.M., Strober, W., et al., 2001. Infections in patients with immunodeficiency with thymoma (Good syndrome). Report of 5 cases and review of the literature. Medicine 80 (2), 123–133.

Case 56

A 42-year-old HIV-positive male is admitted and evaluated for shortness of breath. He is an active smoker with no significant environmental exposures and is non-compliant with his highly active antiretroviral therapy (HAART). A chest X-ray is done showing a moderately large right-sided pleural effusion. A thoracentesis is performed that reveals a slightly bloody exudative pleural effusion. Cytological analysis demonstrates large atypical lymphoid cells with prominent nucleoli and abundant cytoplasm that stained positive for CD20 and human herpesvirus 8 (HHV-8). Chest computed tomography (CT) scan did not reveal any evidence of parenchymal nodules or masses, organ involvement, or lymphadenopathy.

What is the diagnosis?
What are the treatment options?

PRIMARY EFFUSION LYMPHOMA

Primary effusion lymphoma (PEL), also known as body cavity lymphoma, is a unique large B-cell lymphoma that predominately occurs in HIV-positive individuals with a predilection for neoplastic cells to proliferate in the pleural, pericardial, and peritoneal cavities.

The exact etiology behind PEL remains unclear, but it seems to be associated with an underlying viral infection. Studies have shown that HHV-8, or Kaposi's sarcoma-associated herpesvirus, infection plays an important role in disease pathogenesis. The viral infection inhibits cell apoptosis via different gene products (latency-associated nuclear antigen-1 [LANA-1], viral cyclin, and others) resulting in uncontrolled proliferation and neoplastic transformation. Coinfection with Epstein–Barr virus is common in a large majority of patients.

PEL accounts for approximately 3–4% of non-Hodgkin lymphoma (NHL) cases in HIV-positive patients, and around 1% of NHL cases in HIV-negative individuals. The majority of affected HIV-negative patients have some other underlying immunodeficiency. PEL commonly occurs in young males with advanced stage AIDS with no evidence of extranodal lymphoma. The pleural serosal surface is most often involved with patients presenting with symptoms related to the location of the malignant effusion. Pleural involvement results in symptoms of dyspnea, cough, and chest discomfort. Patients with pericardial effusions may have similar symptoms, while those with peritoneal PEL may present with abdominal pain and distention. Patients may also present with joint swelling in some cases.

The effusions are lymphocytic predominant exudates on analysis with evidence of large clonal neoplastic cells on flow cytometry. Diagnosis requires the presence of HHV-8 within the malignant cells. Evidence of HHV-8 infection can be determined by immunohistochemical staining for the LANA-1 latent viral gene product. Chest CT imaging will not reveal any evidence of mass-like lesions, mediastinal adenopathy, or parenchymal pathology in association with the pleural involvement.

PEL is an aggressive, treatment refractory malignancy with a poor prognosis with a median survival of 3–4 months. HIV-negative patients carry a slightly better prognosis than HIV-positive patients with PEL. Patients are often treated with HAART and either EPOCH chemotherapy (cyclophosphamide, doxorubicin, etoposide, vincristine, and prednisone) or CHOP chemotherapy (cyclophosphamide, doxorubicin, vincristine, and prednisone). Rituximab may be added to therapies if cells show CD20 positivity.

TAKEAWAY POINTS

- PEL is a rare NHL seen in HIV-positive patients that only affects the body cavities without mass-like involvement.
- Diagnosis requires the presence of HHV-8 positivity. Prognosis is poor despite HAART and chemotherapy treatment.

FURTHER READING

Boulanger, E., Agbalika, F., Maarek, O., Daniel, M.T., Grollet, L., Molina, J.M., et al., 2001. A clinical, molecular and cytogenetic study of 12 cases of human herpesvirus 8 associated primary effusion lymphoma in HIV-infected patients. Hematol. J. 2 (3), 172–179.

Nador, R.G., Cesarman, E., Chadburn, A., Dawson, D.B., Ansari, M.Q., Sald, J., et al., 1996. Primary effusion lymphoma: a distinct clinicopathologic entity associated with the Kaposi's sarcoma-associated herpes virus. Blood 88 (2), 645–656.

Simonelli, C., Spina, M., Cinelli, R., Talamini, R., Tedeschi, R., Gloghini, A., et al., 2003. Clinical features and outcome of primary effusion lymphoma in HIV-infected patients: a single-institution study. J. Clin. Oncol. 21 (21), 3948–3954.

Case 57

A 46-year-old female is referred for evaluation for a right lower lobe lung opacity that was incidentally seen on a chest X-ray (CXR) that was performed after she was involved in a motor vehicle accident. She is asymptomatic with no complaints of shortness of breath, chest pain, cough, palpitations, or syncope. Her physical examination is unremarkable. A contrasted chest computed tomography (CT) scan performed is shown in Fig. 57.1.

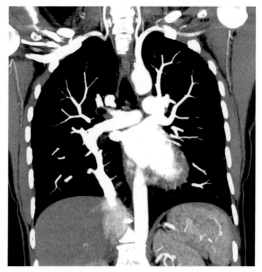

FIGURE 57.1   Contrast enhanced chest computed tomography coronal scan showing an anomalous pulmonary venous return from the right lung pulmonary vein into the inferior vena cava.

What is the diagnosis?
What is Scimitar syndrome?
What is a pseudo-Scimitar syndrome?

PARTIAL ANOMALOUS PULMONARY VENOUS RETURN

Partial anomalous pulmonary venous return (PAPVR), or partial anomalous pulmonary venous connection, is a rare congenital cardiovascular anomaly where some of the pulmonary veins drain into the systemic circulation, commonly at the level of the inferior vena cava, instead of the left atrium. If all the pulmonary veins are involved, it is termed total anomalous pulmonary venous return, a congenital cardiovascular anomaly that is detected early in infancy and is incompatible with life in the absence of a right to left shunting process.

PAPVR most frequently involves the left upper lobe vein. It is seen commonly in association with atrial septal defect (ASD) and less commonly with an atrioventricular defect. Patients may be asymptomatic or present with symptoms related to the degree of left to right shunting such as shortness of breath, chest pain, and palpitations and may even progress to develop pulmonary hypertension and right-sided heart failure. Symptomatic patients will require cardiac catheterization in order to determine the degree of pulmonary to systemic blood flow (Qp:Qs) shunting. Treatment is surgical by correcting the ASD and reimplanting the anomalous vein into the left atrium.

SCIMITAR SYNDROME

The combination of PAPVR, right hypoplasia, and cardiac dextroposition is known as Scimitar syndrome. It is also referred to as pulmonary venolobar syndrome or hypogenetic lung syndrome. PAPVR in this syndrome is characterized by partial venous drainage from the right lung to the inferior vena cava. Patients also have anomalous systemic arterial blood supply from the aorta to the right lung. The name is derived from the shape of a Turkish sword with a curved blade, given the curvilinear pattern seen by the pulmonary vein adjacent to right side of the heart and draining into the inferior vena cava on chest imaging (scimitar sign). The condition may also be associated with pulmonary artery hypoplasia, pulmonary sequestration, horseshoe lung, and accessory diaphragm.

Like in other PAPVR cases, patients may be asymptomatic or present with symptoms of shortness of breath, pulmonary hypertension, and right heart failure due to increased left to right shunting. Patients may also present with recurrent episodes of pneumonia or hemoptysis if associated with pulmonary sequestration. A CXR may reveal curvilinear appearing density, but in some cases this may not be evident due to it being obscured by the right heart border. Diagnosis can be established on chest CT angiography, magnetic resonance imaging angiography, or echocardiography. Treatment of symptomatic patients with increase shunting is by surgical correction.

PSEUDO-SCIMITAR SYNDROME

Pseudo-Scimitar syndrome, or meandering right pulmonary vein, is the incidental finding of an anomalous scimitar vein on CXR that on contrasted chest CT or angiography is found to be draining normally into the left atrium. These patients are usually asymptomatic due to absence of left to right shunting, and there is no need for treatment.

TAKEAWAY POINTS

- Patients with PAPVR may be asymptomatic or present as a result of left to right shunting, pulmonary hypertension, and right heart failure.
- Classic Scimitar syndrome is almost always right sided and is associated with lung hypoplasia and dextroposition of the heart.
- It is important to determine where the anomalous pulmonary vein terminates in order to differentiate meandering right pulmonary vein from classic Scimitar syndrome and its other variants.

FURTHER READING

Ho, M.L., Bhalla, S., Bierhals, A., Gutierrez, F., 2009. MDCT of partial anomalous pulmonary venous return (PAPVR) in adults. J. Thorac. Imaging 24 (2), 89–95.

Rodrigues, M.A., Ritchie, G., Murchison, J.T., 2013. Incidental meandering right pulmonary vein, literature review and proposed nomenclature revision. World J. Radiol. 5 (5), 215–219.

Sanger, P.W., Taylor, F.H., Robicsek, F., 1963. The "scimitar syndrome". Diagnosis and treatment. Arch. Surg. 86, 580–587.

Schramel, F.M., Westermann, C.J., Knaepen, P.J., van den Bosch, J.M., 1995. The scimitar syndrome: clinical spectrum and surgical treatment. Eur. Respir. J. 8 (2), 196–201.

Case 58

A 43-year-old male presents with progressive shortness of breath on exertion over the last year. He has a medical history of hypertension only. He denies any leg swelling, hemoptysis, coughing, fevers, night sweats, or chest pain. His chest X-ray is normal and a chest computed tomography (CT) scan shows an enlarged pulmonary artery but no parenchymal or pleural abnormalities.

An echocardiogram (ECHO) is done showing a normal ejection fraction of 60%, elevated right ventricle systolic pressures, and a hypertrophied and dilated right ventricle. The patient undergoes a right heart catheterization (RHC) that reveals pulmonary hemodynamics consistent with precapillary pulmonary hypertension (PH). As part of his PH workup, he undergoes a lung ventilation-perfusion (VP) scan (Fig. 58.1).

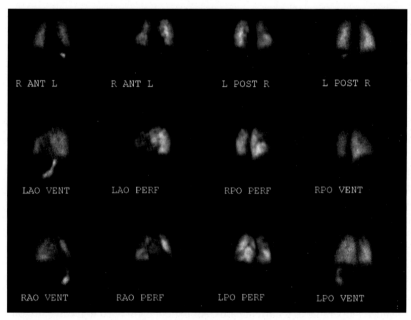

FIGURE 58.1 Ventilation-perfusion scan showing mismatched perfusion defects consistent with CTEPH. *ANT*, anterior; *CTEPH*, chronic thromboembolic pulmonary hypertension; *L*, left; *LAO*, left anterior oblique; *LPO*, left posterior oblique; *PERF*, perfusion; *POST*, posterior; *R*, right; *RAO*, right anterior oblique; *RPO*, right posterior oblique; *VENT*, ventilation.

What is the first-line treatment option for his condition?

CHRONIC THROMBOEMBOLIC PULMONARY HYPERTENSION

Chronic thromboembolic pulmonary hypertension (CTEPH) is a subtype of PH (WHO Group IV) that occurs as a result of an unresolved pulmonary thromboembolism causing persistent obstruction of the pulmonary vessels, progressive pulmonary artery remodeling, and PH, that if untreated will lead to right heart failure and death. The incidence of symptomatic CTEPH after an acute pulmonary embolism is believed to be 3.8%. However, 40–50% of patients with CTEPH will not have a history of pulmonary embolism or deep vein thrombosis.

Patients with CTEPH have a similar clinical presentation to patients with pulmonary arterial hypertension (PAH). Symptomatic patients may complain of progressive dyspnea on exertion, lower extremity swelling, fatigue, and syncope. Elevated right ventricle systolic pressures are present on ECHO, which may also reveal signs of PH and right heart strain such as right ventricle hypertrophy and dilation, and right atrial dilation. Pulmonary function testing findings of isolated reduction in diffusion capacity of the lungs for carbon monoxide may be present. On chest CT imaging, the main pulmonary artery may be dilated and parenchymal findings of mosaic attenuation may be present.

Patients with a clinical presentation and ECHO suggestive of PH should undergo a RHC. RHC hemodynamics in CTEPH are identical to PAH and consistent with precapillary PH ($mPAP \geq 25\,mmHg$, $PAOP \leq 15\,mmHg$). To differentiate CTEPH from PAH, all patients should undergo a VQ lung scan as part of their workup. Multiple mismatched perfusion defects are suggestive of CTEPH. Patients with positive or indeterminate VQ scans should undergo a CT pulmonary angiogram that serves to confirm the presence of CTEPH and evaluate surgical operability.

First-line treatment option for patients who are surgical candidates is pulmonary endarterectomy at an experienced center. This curative procedure, when done at a specialized center, carries a low mortality risk (<4%). For patients who are not surgical candidates, those awaiting surgery, or individuals with persistent or recurrent PH after pulmonary endarterectomy, medical management is directed with PAH-specific therapy. Riociguat, a soluble guanylate cyclase stimulator, is the only approved drug for CTEPH and has been shown to improve 6-min walk distance and functional class. Percutaneous transluminal pulmonary angioplasty with dilation of the narrowed pulmonary arteries is emerging as a treatment option for nonsurgical patients. Lung transplantation is reserved for advanced inoperable or recurrent cases. All patients will need to receive lifelong anticoagulation therapy, generally with warfarin with an INR goal of 2–3.

TAKEAWAY POINTS

- Up to half of patients with CTEPH do not have a history of thromboembolism.
- Diagnosis is established in PH patients with positive VQ scan.
- First-line treatment is pulmonary endarterectomy at an experienced center. Other treatment options include riociguat and percutaneous transluminal pulmonary angioplasty. All patients will need to receive lifelong anticoagulation.

FURTHER READING

Fedullo, P.F., Auger, W.R., Kerr, K.M., Rubin, L.J., 2001. Chronic thromboembolic pulmonary hypertension. N. Engl. J. Med. 345 (20), 1465–1472.

Ghofrani, H.A., D'Armini, A.M., Grimminger, F., Hoeper, M.M., Jansa, P., Kim, N.H., et al., 2013. Riociguat for the treatment of chronic thromboembolic pulmonary hypertension. N. Engl. J. Med. 369 (4), 319–329.

Kim, N.H., Delcroix, M., Jenkins, D.P., Channick, R., Dartevelle, P., Jansa, P., et al., 2013. Chronic thromboembolic pulmonary hypertension. J. Am. Coll. Cardiol. 62 (25 Suppl), D92–D99.

Case 59

A 23-year-old female with a recent history of recurrent right lung pneumonia is referred for further workup after a chest computed tomography (CT) scan was performed to determine the etiology of her recurrent pneumonias. She has been treated as an outpatient with oral antibiotics for pneumonia on three different occasions in the last 6 months. She has no significant past medical history of note, is a nonsmoker with no history of recreational drug use, and with no relevant medical family history. Her chest CT is shown in Fig. 59.1.

A complete workup for immunodeficiencies, cystic fibrosis, alpha-1 antitrypsin deficiency, and connective tissue disorders is unrevealing. She undergoes a video-assisted thoracoscopic surgery right lung biopsy, pathology revealing multiple large cysts with a pseudostratified ciliated columnar epithelium lining.

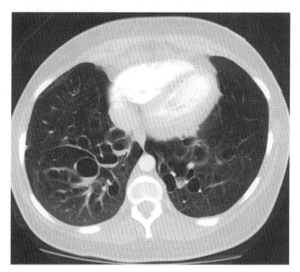

FIGURE 59.1 Chest computed tomography scan showing multiple bilateral lower lobe pulmonary cysts with air fluid levels of varying sizes.

What is the diagnosis?
How should this patient be managed?

CONGENITAL PULMONARY AIRWAY MALFORMATION

Previously referred to as congenital cystic adenomatoid malformation, congenital pulmonary airway malformation (CPAM) is a developmental pulmonary anomaly characterized by the formation of single or multiple cystic lesions as a result of aberrant bronchoalveolar development. It is usually recognized on prenatal ultrasound or early in infancy; however, in some cases it may be diagnosed until adulthood.

The aberrant bronchial development in CPAM is thought to be due to hamartomatous proliferation and cystic dilatation of the terminal respiratory units that end up replacing normal alveoli.

There are five different types:

- Type I: The most common type of CPAM, occurring in 50–70% of cases. It consists of large cysts (>2 cm) that may be surrounded by smaller cysts. It carries the best prognosis among the five types.
- Type II: The second most common form (15–20% of cases). Small cysts (<2 cm) are present and it is often associated with other congenital disorders (cardiac anomalies, pulmonary sequestration, and renal agenesis).
- Type III: Microcysts (<5 mm) are present that involve an entire lobe. These may appear as a consolidation on chest imaging.
- Type IV: It is characterized by unlined cysts that often affect a single lobe.
- Type 0: Also termed acinar dysplasia, it is the least common type of CPAM where the lungs fail to develop.

Adult patients may be asymptomatic or present with symptoms of recurrent infections, pneumothorax, shortness of breath, or hemoptysis. The cystic lesions are often unilateral and involve a single lobe. Chest imaging will be variable depending on the type of CPAM, but most often there are cystic lesions of varying sizes. Lesions may change in size over time due to cystic communications with alveolar structures allowing collateral ventilation.

CPAM cases presenting in adulthood are usually either type I or type II; with only one case of type IV reported in an adult. Pathological examination of lung tissue can help define the type of CPAM. Type I consists of single or multiple large cysts of varying sizes, with a pseudostratified ciliated columnar epithelium lining. Type II is characterized by multiple, uniform small cysts with a cuboidal to columnar epithelium lining.

Treatment of symptomatic patients involves surgical removal of the affected lung, often by a lobectomy. Patients with type I CPAM are at increased risk of bronchoalveolar carcinoma so they should be followed up for its development even after surgery. Asymptomatic patients should be observed clinically. It is unclear if prophylactic resection of the lesions in asymptomatic patients offers any benefit.

TAKEAWAY POINTS

- There are five different types of CPAM, but only two types (I and II) usually manifest in adulthood.
- Symptomatic patients should undergo surgical resection while asymptomatic individuals may be followed clinically.

FURTHER READING

Chin, K.Y., Tang, M.Y., 1949. Congenital adenomatoid malformation of one lobe of a lung with general anasarca. Arch. Pathol. (Chic) 48 (3), 221–229.

Oh, B.J., Lee, J.S., Kim, J.S., Lim, C.M., Koh, Y., 2006. Congenital cystic adenomatoid malformation of the lung in adults: clinical and CT evaluation of seven patients. Respirology 11 (4), 496–501.

Case 60

A 54-year-old man presents with symptoms of shortness of breath for the last 3 months. He states that he has also been experiencing increasing difficulty swallowing and joint pain, and reports that his fingers turn blue on exposure to the cold. He worked as a sandblaster for many years and has a known history of pulmonary silicosis. His chest computed tomography scan is shown in Fig. 60.1 and is unchanged from a scan done a few years prior.

On examination, his skin is taught with loss of the ridges around his knuckles and he is noted to have microstomia. Pulmonary function testing is consistent with an intrinsic restrictive ventilatory defect. An autoimmune work reveals positive antinuclear antibody (ANA; 1:640) in a speckled pattern and a positive anti-Scl-70 antibody. Echocardiography reveals new findings of elevated right ventricular systolic pressure and right ventricle dilatation and hypertrophy.

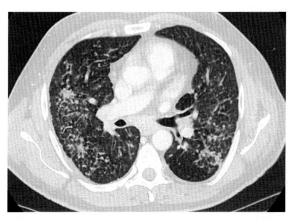

FIGURE 60.1   Chest computed tomography scan showing numerous nodules throughout the lungs with upper lobe predominance and areas of confluence. Significant mediastinal adenopathy is also present.

What is the association between the silica exposure and his current presentation?

ERASMUS SYNDROME

Silicosis is the most common type of pneumoconiosis associated with the inhalation of silica containing mineral dust. Silica exposure has been associated with the development of lung adenocarcinoma, tuberculosis, and certain autoimmune disorders such as systemic sclerosis, rheumatoid arthritis, systemic lupus erythematosus, sarcoidosis, vasculitis, and pemphigus vulgaris.

Erasmus syndrome is defined by the development of systemic sclerosis in a patient who has a history of silica exposure, with or without the development of pulmonary silicosis. It is predominately seen in men with a history of occupational exposure to silica such as miners and sandblasters. The association was first recognized by Erasmus in 1957 in gold miners.

Systemic symptoms and signs of systemic sclerosis (Raynaud's phenomenon, skin sclerosis, fatigue, muscle weakness, and sclerodactyly) often manifest 15 years after the exposure to silica. Autoimmune testing will show ANA positivity and the presence of anti-Scl-70 and/or anti-topoisomerase I antibodies. The diagnosis is established based on the history of exposure to silica, regardless of whether or not pulmonary silicosis is present, and the later development of systemic sclerosis.

The process behind the development of an autoimmune disorder may be related to an abnormal inflammatory response as a result of increased expression of anti-apoptotic proteins in the serum of patients with chronic silica exposure. This leads to an aberrant immune response and dysregulated lymphocyte apoptosis, which results in the development of autoantibodies and the development of systemic sclerosis.

There are no specific guidelines for the treatment of patients with Erasmus syndrome. Patients should avoid exposure to silica if they have not done so already. The systemic sclerosis is treated with corticosteroids and other immunosuppressive agents depending on the severity of the disease. In this case, the patient's parenchymal abnormalities were stable and the worsening shortness of breath was attributed to the development of pulmonary arterial hypertension, which was confirmed on right heart catheterization.

TAKEAWAY POINTS

- Erasmus syndrome is the association between silica exposure and the development of systemic sclerosis.

FURTHER READING

Erasmus, L.D., 1957. Scleroderma in goldminers on the Witwatersrand with particular reference to pulmonary manifestations. S. Afr. J. Lab. Clin. Med. 3, 209–231.

McCormic, Z.D., Khuder, S.S., Aryal, B.K., Ames, A.L., Khuder, S.A., 2010. Occupational silica exposure as a risk factor for scleroderma: a meta-analysis. Int. Arch. Occup. Environ. Health 83, 763–769.

Rustin, M.H., Bull, H.A., Ziegler, V., Mehlhorn, J., Haustein, U.F., Maddison, P.J., et al., 1990. Silica-associated systemic sclerosis is clinically, serologically and immunologically indistinguishable from idiopathic systemic sclerosis. Br. J. Dermatol. 123, 725–734.

Index

'*Note*: Page numbers followed by "f" indicate figures, "t" indicate tables.'

A

Abdominal examination, 15
Abdominal surgery, 66
Actin, 43, 45
Acute eosinophilic pneumonia (AEP), 39
Acute fibrinous and organizing pneumonia (AFOP), 87
Adenovirus infection, 30
Albinism, 115–116
Allergic bronchopulmonary mycosis, 40
Amyloid adenopathy, 2
Amyloidosis, 2, 144
Angiitis, 155
Angiocentric lymphoma. *See* Lymphomatoid granulomatosis (LYG)
Angioedema, 5–6
Angiofollicular lymph node hyperplasia. *See* Castleman disease
Anti-GBM disease, 52
Antigens, 112
Anti-Jo-1, 62–63
Antisynthetase syndrome, 62
Apple-green birefringence, 2–3
Asthma, 56
Atrial septal defect (ASD), 208
Autoimmune disease, 136
Autoimmune pulmonary-renal syndrome, 52

B

β-agonists, 90
Behcet disease, 120
Benign fibrous tumor. *See* Solitary fibrous tumor of the pleura (SFTP)
Benign metastasizing leiomyoma (BML), 45
Bilateral pleural effusions, 35–36, 35f
contarini syndrome, 36
Biliary-pleural fistula, 66
Bilious pleural effusion, 66
Bilothorax, 66
Biopsy
alveolar spaces and interalveolar walls, silicone vacuoles in, 197–198
cervical lymph nodes, 193
dense subpleural fibrosis, 129f
lymphocytic bronchiolitis, 58f
organizing pneumonia, 58f
skin rash, 26
Birbeck granules, 27
Birt-Hogg-Dubé (BHD) syndrome, 48
Blunt trauma, 84
Bone cement implantation syndrome (BCIS), 122
Brett syndrome/Swyer-James-MacLeod, 30
Bronchiectasis, 11, 158–159
Bronchoalveolar lavage (BAL), 22, 51
Bronchoscopy, 37
Bullectomy, 81

C

C1q precipitin antibody, 6
CA-125 levels, 16
Calcified spherules, 162
Castleman disease
distinct histological subtypes, 68
MCD, 68–69
UCD variant, 68–69
Castleman- Kojima disease, 68
Catamenial hemoptysis, 18
Catamenial hemothorax, 18–19
Catamenial pneumothorax, 18
Cavity, 84
CD1a cells, 26, 26f
CD20+ B cells, 153
CD3+ T cells, 153
Celiac disease, 133–134
Ceroid lipofuscin, 116
Chest computed tomography (CT) scan, 1, 1f
apical irregular cysts and nodules, 25, 25f
bilateral bullous disease, 80f
bilateral pulmonary nodules, 153f
bronchiectasis, 29f
crazy-paving pattern, 23
dilated trachea and posterior tracheal diverticuli, 157f
fat density left lower lobe, 149f
ground-glass opacity, 21–22, 21f

Chest computed tomography (CT) scan
  (Continued)
  hyperlucent lung, 29f, 30
  left lower lobe, thin-walled cysts in, 143f
  left-sided pleural effusion, 185f
  lower lobes, dextrocardia and bronchiectasis
    of, 182f
  mediastinal adenopathy, 67
  right upper, bullous lesion in, 31, 31f
  subpleural micronodular lesions, 177f
Chest X-ray (CXR), 21
  bilateral diffuse pulmonary opacities, 37, 37f
  dextrocardia and right gastric bubble, 181f
  micronodules, 161f
  nodular lesions, 43f
Cholecystopleural fistula, 66
Cholethorax, 66
Chronic eosinophilic pneumonia (CEP), 39
Chronic silicone embolism syndrome, 198
Chronic thromboembolic pulmonary
    hypertension (CTEPH), 212
Churg Strauss disease, 40
Chyliform/cholesterol effusion. *See*
    Pseudochylothorax
Chylomicrons, 12
Chylopericardium, 72
Chylothorax, 11–12
  management of, 12
  pseudochylothorax effusions, 12
  thoracic bone involvement, 72
  thoracic duct, 11
Chylous effusions, 11
Chylous pleural effusions, 11
Cidofovir, 175
Common variable immunodeficiency
    (CVID), 57
Complete tracheal rings, 56
Congenital cystic adenomatoid malformation,
    216
Congenital pulmonary airway malformation
    (CPAM)
  bronchial development in, 216
  different types, 216
  treatment, 216
Congenital tracheal anomalies, 56
Congo red stain, 1f, 2–3
Contarini syndrome, 36
Corticosteroids, 59
Creatinine, 110
Cryoglobulinemia-associated diffuse alveolar
    hemorrhage (DAH), 94
CTEPH. *See* Chronic thromboembolic
    pulmonary hypertension (CTEPH)

Cystic lung diseases, 48
Cytoplasmic vacuoles, 150

D
DAH. *See* Diffuse alveolar hemorrhage (DAH)
Deep vein thrombosis, 120
Dendriform, 178
Dense fibrovascular core, polypoid tissue with,
    173, 174f
Desmin, 43, 45
Diaphragm, 90
Diffuse alveolar damage, 87
Diffuse alveolar hemorrhage (DAH),
    51, 93–94
Diffuse idiopathic pulmonary neuroendocrine
    cell hyperplasia (DIPNECH), 102
Diffuse interstitial form, 2
Diffuse pulmonary ossification (DPO)
  dendriform, 178
  diagnosis of, 178
  low calcium diets and systemic steroid, 178
  lung biopsy, 178
  nodular, 178
  types of, 178
Doege–Potter syndrome, 76

E
ECD. *See* Erdheim-Chester disease (ECD)
Echocardiography, 15
Elastin, 131
Emperipolesis, 193–194
Emphysema, 5f, 6
Endobronchial ultrasound, 1
Endometrial lung nodules, 19
Epstein-Barr virus (EBV), 153, 155, 204
Erasmus syndrome, 220
Erdheim-Chester disease (ECD), 106
Extracorporeal shock wave lithotripsy, 109
Extrinsic allergic alveolitis, 112

F
Fat embolism syndrome, 122–123
Fever, 135–136
Fibrin, 87
Fibrofolliculomas, 48
Fibrosis pattern, 126
Fire-eaters, 150
FLCN gene, 48
Foamy macrophages, 149–150
Follicular bronchiolitis, 59
Folliculin protein, 48

G

Giant bullae, 81
Glomerulonephritis, 94
Gluten-free diet, 134
Good syndrome, 200
Goodpasture syndrome, 52
Gorham-Stout syndrome, 72
Graft-versus-host disease (GVHD), 148
Granulocyte macrophage colony-stimulating
   factor (GM-CSF), 22–23
Granulomatous organizing
   pneumonia, 87
Granulomatous-lymphocytic interstitial lung
   disease (GLILD), 59
   common symptoms of, 59
   CVID, 59
   etiology of, 59

H

Hairy kidneys, 106
Heart failure, 35–36
Hemangiomatosis, 72
Hematopoietic stem cell transplantation, 148
Hemodialysis, 52
Hepatitis C virus, 94
Hepatopulmonary syndrome, 166
Hereditary hemorrhagic telangiectasia
   (HHT), 167
Hermansky-Pudlak syndrome (HPS), 116
Highly active antiretroviral therapy (HAART),
   203–204
High-resolution computed tomography
   (HRCT) scan, 61
Histiocytic necrotizing lymphadenitis, 136
Histiocytosis X, 26–27
Hornstein-Knickenberg syndrome, 48
Hughes-Stovin syndrome, 120
Human herpesvirus 8 (HHV-8), 204
Human papillomavirus (HPV)
   infection, 175
Hydrocarbons, 52–53
Hypersensitivity pneumonitis (HP)
   corticosteroids, treatment with, 114
   etiology of, 112
   symptoms, 112
   types of, 112, 113t
Hypertrophic pulmonary osteopathy, 76
Hypocomplementemic urticarial vasculitis
   syndrome (HUVS), 6
Hypogammaglobulinemia, 59, 200
Hypoglycemia, 76
Hypoxia, 121

I

Idiopathic giant bullous emphysema, 81
Idiopathic hypereosinophilic syndrome
   (IHES), 40
Idiopathic pleuroparenchymal fibroelastosis
   (IPPFE), 131
Idiopathic pulmonary hemosiderosis (IPH),
   134
IgG4-related sclerosing disease, 126
IgG4-related systemic disease, 126
Immotile-cilia syndrome, 183
Immunodeficiency, 200
Immunoglobulin A gammapathy, 134
Immunosuppressive therapy, 6, 52, 87
Interleukin-6 (IL-6), 68–69
Interstitial lung disease, 62–63
Intravenous immunoglobulin (IVIG) therapy,
   200
IPPFE. *See* Idiopathic pleuroparenchymal
   fibroelastosis (IPPFE)

K

Kaposi's sarcoma-associated herpesvirus,
   204
Kartagener syndrome, 183
Kikuchi–Fujimoto disease (KFD), 136

L

Lane-Hamilton syndrome, 134
Langerhans cell histiocytosis (LCH)
   clinical manifestations, 27
   cystic lung disease, differential diagnosis
      of, 26
   electron microscopy, 27
   pulmonary function testing, 27
   S100, CD207, and CD1a, 27
   skin biopsy, 26, 26f
   treatment strategies, 27
Latency-associated nuclear antigen-1
   (LANA-1), 204
Light chain deposition disease
   (LCDD), 144
Lipid-laden macrophages, 150
Lipoid pneumonia
   acute/chronic, 150
   diagnosis, 150
   endogenous, 150
   exogenous, 150
   treatment, 151
Liver function test, 15
Löffler syndrome, 40

Lung disease
  AEP, 39
  CIVD, 59
  cystic, differential diagnosis of, 26
  GLILD, 59
  PAP, 22–23
  papillary structures, 31, 32f
  placental transmogrification of, 33
Lung transplantation, 27, 212
  GVHD, 148
Lymphadenitis, 136
Lymphadenopathy, 136
Lymphangioleiomyomatosis (LAM), 140
Lymphangiomas, 141
Lymphangiomatosis, 72, 141
Lymphatic dysplasia syndrome, 141
Lymphedema, 11
Lymphocytic interstitial pneumonia (LIP), 59
Lymphoid hyperplasia, 59
Lymphomatoid granulomatosis (LYG), 155
Lymphoplasmacytic tissue, 126
Lymphoproliferative disorder, 68–69
  EBV, 155–156
Lysosomes, 116

M
Maculopapular rash, 147
MAGIC syndrome, 190
Massive bone osteolysis, 72
Meandering right pulmonary vein, 209
Mechanic's hands, 62
Mechanistic target of rapamycin (mTOR), 140
Mediastinal adenopathy, 1
Medium-chained triglycerides, 12
Methyl methacrylate (MMA), 122
Mounier-Kuhn syndrome
  pulmonary function testing, 158
  treatment, 158–159
  types of, 158
  WCS, 159
Multicentric Castleman disease (MCD), 68
  HIV and HHV-8 infection, 68–69
Multiple myeloma, 185–186
Myelomatous pleural effusion (MPE), 186
Myositis, 62

N
Nd–YAG laser, 2
Neuroendocrine cell hyperplasia, 101
Nodular form, 2, 178
Non-Hodgkin lymphoma (NHL), 204
Non-Langerhans histiocytes, 106–107

O
Obliterative bronchiolitis, 30
Octreotide, 12
Osler-Weber-Rendu syndrome, 167
Ovarian tumor, 16

P
PAM. *See* Pulmonary alveolar microlithiasis
    (PAM)
Pancytopenia, 147
Parapneumonic effusion, 35–36
Paraseptal emphysema, 81
Parasitic infections, 40
Partial anomalous pulmonary venous
    connection, 208
Partial anomalous pulmonary venous return
    (PAPVR), 208
  Pseudo-Scimitar syndrome, 209
  Scimitar syndrome, 208
PCD. *See* Primary ciliary dyskinesia (PCD)
PEL. *See* Primary effusion lymphoma (PEL)
Percutaneous transcatheter embolization,
    166–167
Peribronchovascular lymphatic dilation, 140
Pericardial effusions, 11
PFTs. *See* Pulmonary function tests (PFTs)
Pierre-Marie-Bamberger syndrome, 76
Placental transmogrification/placentoid bullous
    lesion, 33
Plasmapheresis, 52
Pleural effusion, 11, 110
  bilothorax, 66
Pleural fluid, 9, 10f
Pleurodesis, 110, 186
Pleuroparenchymal fibroelastosis (PPFE), 131
PNEC. *See* Pulmonary neuroendocrine cell
    (PNEC)
Pneumoconiosis, 220
Pneumothorax, 47
  catamenial pneumothorax, 18
  TPP, 84
  VLS, 81
Polyangiitis, eosinophilic granulmatosis with,
    40
Polydimethylsiloxane, 198
Polymerase chain reaction (PCR), 147
Polypoid intraluminal lesion, 173f
PPFE. *See* Pleuroparenchymal fibroelastosis
    (PPFE)
Primary ciliary dyskinesia (PCD), 183
Primary effusion lymphoma (PEL), 204
Pseudochylothorax, 12

Pseudo-pseudo-Meigs syndrome, 16
Pseudo-Scimitar syndrome, 209
Pseudotumor lesion, 106
Pulmonary alveolar microlithiasis (PAM), 162
Pulmonary alveolar proteinosis (PAP)
    crazy-paving pattern, 22
    GM-CSF therapy, 23
    symptoms, 23
    transbronchial biopsies, 22
    whole lung lavage, 23
Pulmonary amyloidosis, 2
Pulmonary arterial hypertension (PAH), 97–99, 212
Pulmonary arteriovenous malformations (PAVM)
    causes of, 166
    chest CT scan, 166
    echocardiography, 166
    HHT, 167
    percutaneous transcatheter embolization, 166–167
    symptoms, 166
    treatment, 166–167
Pulmonary artery aneurysm rupture, 120
Pulmonary artery occlusion pressure (PAOP), 97
Pulmonary capillary hemangiomatosis (PCH), 99
Pulmonary cysts, 48
Pulmonary embolism, 121
    CTEPH, 212
Pulmonary eosinophilia, 40
    etiologies of, 39–40
Pulmonary fibrosis, 116
Pulmonary function tests (PFTs), 5, 21, 27
    restrictive ventilatory defect, 57, 59, 87
    obstructive ventilatory defect, 87
    intrinsic restrictive ventilatory defect, 115, 129
Pulmonary hypertension (PH), 170, 211
Pulmonary light chain deposition disease (LCDD), 144
Pulmonary lymphangiectasis, 141
Pulmonary lymphomatoid granulomatosis, 155
Pulmonary neuroendocrine cell (PNEC), 102
Pulmonary nodules, 102
Pulmonary ossification, 178
Pulmonary tumor emboli syndrome, 170
Pulmonary tumor thrombotic microangiopathy (PTTM), 170
Pulmonary vascular resistance (PVR), 97
Pulmonary veno-occlusive disease (PVOD)
    CT chest scan, 98
    echocardiography, 98
    etiology of, 98
    PAH, 98–99
    symptoms, 98
Pulmonary-renal syndromes, 52

R
Rapamycin, 140
Recurrent respiratory papillomatosis (RRP), 175
Relapsing polychondritis (RP)
    bronchial stenosis, 191
    chest CT imaging, 190–191
    clinical diagnosis, 190
    criteria, 190
    ear cartilage, 190
    laryngeal chondritis, 190
    nasal cartilage, 190
Renal angiomyolipomas, 140
Renal tumors, 48
    BHD, 48
Retrograde menstruation, 18
Retroperitoneal fibrosis, 106
Right heart catheterization (RHC), 97, 211
Rosai-Dorfman disease (RDD), 194
RP. See Relapsing polychondritis (RP)
RRP. See Recurrent respiratory papillomatosis (RRP)

S
Sandstorm lung, 162
Scimitar syndrome, 208
Scimitar vein, 209
Shrinking lung syndrome (SLS)
    corticosteroids, use of, 90
    etiology of, 90
    physical examination, 90
    SLE, pulmonary manifestations of, 90
Silica, 220
Silicone embolism syndrome, 198
    chronic, 198
Silicosis, 220
Sinus histiocytosis with massive lymphadenopathy, 194
Sinus symptoms, 11
Sirolimus, 140–141
Situs inversus, 183
Skin lesions, 48
SLC34A2 gene, 162
SLE. See Systemic lupus erythematosus (SLE)
Smooth muscle cells, 140–141

Solitary fibrous tumor of the pleura (SFTP)
CT scan, 76
immunohistochemical staining, 76
paraneoplastic syndromes, 76
prognosis of, 77
Squamous cell carcinoma, 175
Storiform/whorled fibrosis pattern, 126
Stove-pipe trachea, 56
Swyer-James syndrome (SJS), 30
Systemic lupus erythematosus (SLE), 6, 15, 89–90
CA-125 levels, 16
KFD, 136
pleural effusions, causes for, 16
Systemic sclerosis, 220

T
Theophylline, 90
Thoracentesis, 9
Thoracic endometriosis
catamenial hemoptysis, 18
catamenial pneumothorax, 18
endometrial lung nodules, 19
manifestations of, 18
retrograde menstruation, 18
Thoracobilia, 66
Thymomas, 199
paraneoplastic/autoimmune conditions, 200–201
Tissue biopsy, 2
Tissue factor, 169–170
Tjalma syndrome, 16
Total anomalous pulmonary venous return, 208
Tracheobronchial amyloidosis, 2
Tracheobronchomalacia, 158–159, 190–191
Tracheobronchomegaly, 158
Transaminitis, 147

Transudate, 35
Traumatic pneumatoceles, 84
Traumatic pulmonary pseudocysts (TPP), 84
Tropical pulmonary eosinophilia, 40
Tuberous sclerosis complex (TSC), 140–141
Tumorlets, 102
Type I bullous disease, 81

U
Unicentric Castleman disease (UCD), 68–69
treatment of, 69
Unilateral pulmonary cysts/bullous disease, 33
Urinothorax, 110
Uterine leiomyomas, 45
hysterectomy, 43, 45

V
Vanishing lung syndrome (VLS), 81
Vascular endothelial growth factor (VEGF), 169–170
Ventilation-perfusion (VP) scan, 211
Video-assisted thoracoscopic surgery (VATS), 43, 45
Vimten, 45
Viral cyclin, 204
Viral infection, 136

W
Whole lung lavage, 23
Williams-Campbell syndrome (WCS), 159

Y
Yellow nail syndrome, 11
diagnosis of, 12
thick yellow fingernails, 9, 9f
thick yellow toenails, 9, 9f